Max Planck

Scientific Autobiography and Selected Lectures on Theoretical Physics

Edited by Vesselin Petkov

MINKOWSKI
Institute Press

Max Karl Ernst Ludwig Planck
23 April 1858 – 4 October 1947

Cover: https://commons.wikimedia.org/wiki/File:Max_Planck_
1933.jpg#/media/File:Max_Planck_1933.jpg

ISBN: 978-1-927763-88-9 (softcover)
ISBN: 978-1-927763-89-6 (ebook)

Minkowski Institute Press
Montreal, Quebec, Canada
http://minkowskiinstitute.org/mip/

For information on all Minkowski Institute Press publications
visit our website at http://minkowskiinstitute.org/mip/books/

EDITOR'S PREFACE

This volume includes new publications of Max Planck's "Scientific Autobiography,"[1] eight lectures on theoretical physics delivered by him at Columbia university in the Spring of 1909[2] and his Nobel Prize Address[3] "The Origin and Development of the Quantum Theory," delivered before the Royal Swedish Academy of Sciences at Stockholm on June 2, 1920.

The idea of selecting Planck's works for this volume is simple – to let Planck talk about his research in his "Scientific Autobiography" and then the selected lectures demonstrate his achievements and his views on "The Present System of Theoretical Physics" as he himself described the subject of his lectures.

Due to Planck's crucial support for the fast acceptance of Einstein's special relativity, as we know, I think it is appropriate to say the following regarding Planck's eighth lecture "General Dynamics. Principle of Relativity." As Planck immediately recognized the revolutionary character of Einstein's 1905 paper on special relativity and was instrumental

---

[1]Originally published in Max Planck, *Scientific Autobiography And Other Papers*. Translated by F. Gaynor (Williams & Norgate Ltd, London 1950).

[2]Originally published in Max Planck, *Eight Lectures on Theoretical Physics*, Delivered at Columbia university in 1909. Translated by A.P. Wills (Columbia University Press, New York 1915).

[3]Originally published in Max Planck, *The Origin and Development of the Quantum Theory*. Translated by H. T. Clarke and L. Silberstein (Oxford University Press, Oxford 1922).

i

in its acceptance, it is worth mentioning here how he regarded the concept of relativistic mass, especially in view of the fact that during the last three decades physicists have witnessed (rather endured) "what has probably been the most vigorous campaign ever waged against the concept of relativistic mass."[4]

Not only did Planck not question this new concept in physics, but he followed Einstein's initial[5] introduction of transverse mass and longitudinal mass (see Planck's eighth lecture "General Dynamics. Principle of Relativity" p. 168). It is highly unlikely that a physicist of the rank of Planck would follow blindly Einstein's ideas; that is why it appears certain that Planck's view on the concept of relativistic mass was based on deep understanding of special relativity.

If there were genuine unsurmountable problems with the relativistic mass[6] the concept would have been abandoned immediately after its introduction in 1905 since at that time it was already a great challenge to the understanding of mass. Not only was the concept of relativistic mass not attacked at that time but was later adopted in the textbooks on relativ-

---

[4]M. Jammer, Concepts of Mass in Contemporary Physics and Philosophy (Princeton University Press, Princeton 2000), Ch. 2. Max Jammer devoted a whole chapter to that unprecedented and unjustified campaign. See also the Appendix (by the editor) "On Relativistic Mass" in the new publication of five works by Einstein: A. Einstein, *Relativity*, Edited by V. Petkov (Minkowski Institute Press, Montreal 2018).

[5]Probably we will never know why Einstein simply abandoned the concept of relativistic mass and remained silent about it, mentioning it only once in a 1948 letter to Lincoln Barnett – Einstein's comment (and its two translations) is given in the Appendix "On Relativistic Mass" mentioned above.

[6]There seem to be open questions involved in the relativistic mass (e.g., mass appears to be a tensor) and some colleagues in the last three decades might have decided that it is easier to reject the notion of relativistic mass altogether instead of facing and dealing with the open questions which it revealed (if that is indeed the case, it would hardly be physics at its best).

ity.[7]

But as the features of the external world discovered by physics are not determined by voting, it does not really matter what individual physicists say or what is written even in the best physics textbooks. What does matter is what the Ultimate Judge in physics (the experimental evidence) says and in the case of relativistic mass the verdict, I think, is exceedingly clear: the concept of relativistic mass reflects an experimental fact – since *mass is defined as the measure of the resistance a particle offers to its acceleration* (which is the accepted definition based on the experimental evidence) and since *it is also an experimental fact that a particle's resistance to its acceleration increases* indefinitely (in a given reference frame) as the particle's velocity approaches the speed of light (in the same reference frame), it inescapably follows that the particle's mass increases when its velocity increases.

Planck's works were typeset in LaTeX and noticed typos in the texts and the equations were corrected and several obvious omissions and errors were also corrected.

January 27, 2020   Vesselin Petkov
                                                   Minkowski Institute

---

[7] Here are several of the best textbooks on relativity where the concept of relativistic mass is presented as an integral part of special relativity:
- W. Pauli, *Theory of Relativity*. Translated by G. Field. (Dover Publications, Inc., New York 1958); originally written in 1921
- A. S. Eddington, *The Mathematical Theory of Relativity* (Cambridge University Press 1923)
- C. Møller, *The Theory of Relativity* (Oxford UNiversity Press 1952)
- J. L. Synge, *Relativity: The Special Theory* (Nord-Holand, Amsterdam 1955)
- P. G. Bergmann, *The Special Theory of Relativity*. In: S. Flügge (ed.), Principles of *Electrodynamics and Relativity*, Volume IV of *Encyclopedia of Physics* (Springer, Berlin 1962)
- J. L. Anderson, *Principles of Relativity Physics* (Academic Press, New York 1967)

iv

CONTENTS

v

Part I

Max Planck's Scientific Autobiography

1 A Scientific Autobiography

My original decision to devote myself to science was a direct result of the discovery which has never ceased to fill me with enthusiasm since my early youth – the comprehension of the far from obvious fact that the laws of human reasoning coincide with the laws governing the sequences of the impressions we receive from the world about us; that, therefore, pure reasoning can enable man to gain an insight into the mechanism of the latter. In this connection, it is of paramount importance that the outside world is something independent from man, something absolute, and the quest for the laws which apply to this absolute appeared to me as the most sublime scientific pursuit in life.

These views were bolstered and furthered by the excellent instruction which I received, through many years, in the *Maximilian-Gymnasium* in Munich from my mathematics teacher, Hermann Müller, a middle-aged man with a keen mind and a great sense of humor, a past master at the art of making his pupils visualize and understand the meaning of the laws of physics.

My mind absorbed avidly, like a revelation, the first law I knew to possess absolute, universal validity, independently from all human agency: The principle of the conservation of energy, I shall never forget the graphic story Müller told us, at his raconteur's best, of the bricklayer lifting with great effort a heavy block of stone to the roof of a house. The work he thus performs does not get lost; it remains stored up, perhaps for many years, undiminished and latent in the block of stone, until one day the block is perhaps loosened and drops on the head of some passerby.

After my graduation from the *Maximilian-Gymnasium* I attended the University, first in Munich for three years, then in Berlin for another year. I studied experimental physics and mathematics; there were no professorships or classes in theoretical physics as yet. In Munich, I attended the classes of the physicist Ph. von Joliy, and of the mathematicians Ludwig Seidel and Gustav Bauer. I learned a great deal from these three professors, and I still retain them in reverent memory. But I did not realize until I came to Berlin that in matters concerned with science they had really just a local significance, and it was in Berlin that my scientific horizon widened considerably under the guidance of Hermann von Helmholtz and Gustav Kirchhoff, whose pupils had every opportunity to follow their pioneering activities, known and watched all over the world. I must confess that the lectures of these men netted me no perceptible gain. It was obvious that Helmholtz never prepared his lectures properly. He spoke haltingly, and would interrupt his discourse to look for the necessary data in his small note book; moreover, he repeatedly made mistakes in his calculations at the blackboard, and we had the unmistakable impression that the class bored him at least as much as it did us. Eventually, his classes became more and more deserted, and finally they were attended by only three students; I was one of the three, and my friend, the subsequent astronomer Rudolf Lehmann-Filhés, was another.

Kirchhoff was the very opposite. He would always deliver a carefully prepared lecture, with every phrase well balanced and in its proper place. Not a word too few, not one too many. But it would sound like a memorized text, dry and monotonous. We would admire him, but not what he was saying.

Under such circumstances, my only way to quench my thirst for advanced scientific knowledge was to do my own reading on subjects which interested me; of course, these were the subjects relating to the energy principle. One day, I happened to come across the treatises of Rudolf Clausius, whose lucid style and enlightening clarity of reasoning made an enormous impression on me, and I became deeply absorbed in his articles, with an ever increasing enthusiasm. I appreciated especially his exact formulation of the two Laws of Thermodynamics, and the sharp distinction which he was the first to establish between them. Up to that time, as a consequence of the theory that heat is a substance, the universally accepted view had been that the passing of heat from a higher to a lower temperature was analogous to the sinking of a weight from a higher to a lower position, and it was not easy to overcome this mistaken opinion.

Clausius deduced his proof of the Second Law of Thermodynamics from the hypothesis that *"heat will not pass spontaneously from a colder to a hotter body"* But this hypothesis must be supplemented by a clarifying explanation. For it is meant to express not only that heat will not pass directly from a colder into a warmer body, but also that it is impossible to transmit, by any means, heat from a colder into a hotter body without there remaining in nature some change to serve as compensation.

In my endeavor to clarify this point as fully as possible, I discovered a way to express this hypothesis in a form which I considered to be simpler and more convenient, namely: *"The process of heat conduction cannot be completely reversed by any means."* This expresses the same idea as the wording of Clausius, but without requiring an additional clarifying explanation. A process which in no manner can be completely reversed I called a *"natural"* one. The term for it in universal use today, is: *"Irreversible."*

Yet, it seems impossible to eradicate an error which arises out of an all too narrow interpretation of Clausius' law, an error against which I have fought untiringly all my life. To this very day, instead of the definition I just mentioned, one often finds irreversibility defined as "An irreversible process is one which cannot take place in the opposite direction." This formulation is insufficient. For it is quite possible to conceive of a process which cannot take place in the opposite direction but which can in some fashion be completely reversed.

Since the question whether a process is reversible or irreversible depends solely on the nature of the initial state and of the terminal state of the process, but not on the manner in which the process develops, in the case of an irreversible process the terminal state is in a certain sense more important than the initial state – as if, so to speak, Nature "preferred" it to the latter. I saw a meas- ure of this "preference" in Clausius' entropy; and I found the meaning of the Second Law of Thermodynamics in the principle that in every natural process the sum of the entropies of all bodies involved in the process increases. I worked out these ideas in my doctoral dissertation at the University of Munich, which I completed in 1879.

The effect of my dissertation on the physicists of those days was nil. None of my professors at the University had any understanding for its contents, as I learned for a fact in my conversations with them. They doubtless permitted it to pass as a doctoral dissertation only because they knew me by my other activities in the physical laboratory and in the mathematical seminar. But I found no interest, let alone approval, even among the very physicists who were closely concerned with the topic. Helmholtz probably did not even read my paper at all. Kirchhoff expressly disapproved of its contents, with the comment that the concept of entropy, whose magnitude could be measured by a reversible process only, and therefore was definable, must not be applied to irreversible processes. I did not succeed in reaching Clausius. He did not answer my letters, and I did not find him at home when I tried to see him in person in Bonn. I carried on a correspondence with Carl Neumann, of Leipzig, but it remained totally fruitless.

However, deeply impressed as I was with the importance of my self-imposed task, such experiences could not deter me from continuing my studies of entropy, which| I regarded as next to energy the most important property of physical systems. Since its maximum value indicates a state of equilibrium, all the laws of physical and chemical equilibrium follow from a knowledge of entropy. I worked this out in detail during the following years, in a number of different researches. First, in investigations on the changes in physical state, presented in my probationary paper at Munich in 1880, and later in studies on gas mixtures. All my investigations yielded fruitful results. Unfortunately, however, as I was to learn only subsequently, the very same theorems had been obtained before me, in fact partly in an even more universal form, by the great American theoretical physicist Josiah Willard Gibbs, so that in this particular field no recognition was to be mine.

While an instructor in Munich, I waited for years in vain for an appointment to a professorship. Of course, my prospects for getting one were slight, for theoretical physics had not as yet come to be recognized as a special discipline. All the more compellingly grew in me the desire to win, somehow, a reputation in the field of science.

Guided by this desire, I decided to submit a paper for the prize to be awarded in 1887 by the Philosophical Faculty of Göttingen. The subject to be discussed was, "The Nature of Energy." After I had completed my paper, in the spring of 1885, I was offered the associate professorship in theoretical physics at the University of Kiel. This offer came as a message of deliverance. The moment when I paid my respects to Ministerial Director Althoff in his suite in the Hotel Marienbad, and he informed me of the particulars and conditions of my appointment, was, and will always be, one of the happiest of my life. For even though my life in my parents' house was as beautiful and contented as any man could wish for, my longing for independence kept growing within me, and I was yearning for a home of my own.

To be sure, I suspected, and by no means without reason, that this smile of good fortune was actually not so much a reward for my scientific accomplishments as a practical result of the circumstance that Gustav Karsten, Professor of Physics in Kiel, happened to be a close friend of my father. Nevertheless, this realization could not mar my supreme happiness, and I was firmly resolved to justify the confidence in me in every way in my power.

I soon moved to Kiel, where I put the finishing touches on my paper, and submitted it in Göttingen. It won second prize. Besides my entry, two other papers had been submitted on the subject, but these two were awarded no prize at all. Obviously, I was wondering why my paper had failed to win first prize, and I found the answer in the text of the detailed decision of the Faculty of Göttingen. The judges set forth a few points of criticism of minor import, and then stated : "Finally, the Faculty must withhold its approval from the remarks in which the author tries to appraise Weber's Law." Now, the story behind these remarks was: W. Weber was the Professor of Physics in Göttingen, between whom and Helmholtz there existed at the time a vigorous scientific controversy, in which I had expressly sided with the latter, I think that I make no mistake in considering this circumstance to have been the main reason for the decision of the Faculty of Göttingen to withhold the first prize from me. But while with my attitude I had incurred the displeasure of the scholars of Göttingen, it gained me the benevolent attention of those of Berlin, the results of which I was soon to feel.

No sooner had I finished my paper for Göttingen than

I returned to my favorite subject, and wrote a number of monographs, which I published under the collective title, *On The Principle of the Increase of Entropy*. In these articles I discussed the laws of chemical reactions, of the dissociation of gases, and finally the properties of dilute solutions. With respect to the latter, my theory led to the conclusion that the values of the lowering of the freezing point, observed in many salt solutions, could be explained only by a dissociation of the substances dissolved, and that this finding constituted a thermodynamic foundation for the electrolytic dissociation theory advanced by Svante Arrhenius approximately at the same time. This conclusion, unfortunately, got me into an unpleasant conflict. For Arrhenius challenged, in a rather unfriendly manner, the admissibility of my arguments, pointing out that his theory related to ions, i.e. electrically charged particles, I could reply only that the laws of thermodynamics were valid regardless of whether or not the particles carried a charge.

In the spring of 1889, after the death of Kirchhoff, I accepted the invitation, extended to me upon the recommendation of the Faculty of Philosophy of Berlin, to take his place at the University, to teach theoretical physics. First, I was an associate professor, and from 1892, a full professor. These were the years of the widest expansion of my scientific outlook and way of thinking. For this was the first time that I came in closer contact with the world leaders in scientific research in those days – Helmholtz, above all the others. But I learned to know Helmholtz also as a human being, and to respect him as a man no less than I had always respected him as a scientist. For with his entire personality, integrity of convictions and modesty of character, he was the very incarnation of the dignity and probity of science. These traits of character were supplemented by a true human kindness, which touched my heart deeply. When during a conversation he would look at me with those calm, searching, penetrating, and yet so benign eyes, I would be overwhelmed by a feeling of boundless filial trust and devotion, and I would feel that I could confide in him, without reservation, everything that I had on my mind, knowing that I would find him a fair and tolerant judge; and a single word of approval, let alone praise, from his lips would make me as happy as any worldly triumph.

I had this experience on several occasions. One of them was when he thanked me emphatically after my memorial address on Heinrich Hertz, delivered before the Physical Society; another, when he expressed his agreement with my theory of chemical solutions, shortly before my election to the Prussian Academy of Sciences. I shall treasure the memory of every one of these thrilling moments to the end of my days.

Besides Helmholtz, I was soon on amicable terms with Wilhelm von Bezold, whom I had known from Munich. Likewise, with August Kundt, the temperamental Director of the Physical Institute, universally liked for his genuine kind human feelings.

The other physicists were not so easy to approach. There was, for instance, Adolph Paalzow, the physicist of the School of Engineering of Charlottenburg, a gifted experimenter, and a typical Berliner. He would always treat me cordially, yet always make me feel that he had really not much use for me. In those days, I was the only theorist, a physicist *sui generis*, as it were, and this circumstance did not make my *debut* so easy. Also, I had a distinct feeling that the instructors at the Physical Institute were politely but clearly trying to keep me at arm's length. But in the course of time, as we got better acquainted, our relationship assumed a friendlier aspect; one of them, Heinrich Rubens, eventually became my close personal friend, and our friendship was ended only by his death, at an all too early age.

By a sheer whim of fate, no sooner had I reported to my post in Berlin than I was temporarily assigned a task in a field quite remote from my self-chosen special branch of physics. Just at that time, the Institute for Theoretical Physics happened to receive a large harmonium, of pure untempered tuning, a product of the genius of Carl Eitz, a public school teacher in Eisleben, built by the Schiedmayer piano factory of Stuttgart for the Ministry. I was given die task of using this musical instrument for a study of the untempered, "natural" scale. I delved into the problem with keen interest, in particular with regard to the question concerning the part played by the "natural" scale in our modern vocal music without instrumental accompaniment. These studies brought me the discovery, unsuspected to a certain degree, that the tempered scale was positively more pleasing to the human ear, under all circumstances, than the "natural," untempered scale. Even in a harmonic major triad, the natural third sounds feeble and inexpressive in comparison with the tempered third. Indubitably, this fact can be ascribed ultimately to a habituation through years and generations. For before Johann Sebastian Bach, the tempered scale had not been at all universally known.

My removal to Berlin not only enabled me to associate with interesting personages, but also brought about a sizable expansion of my scientific correspondence. First of all, I became interested in the extremely fruitful theory formulated by W. Nernst, of Göttingen. According to this theory, the electric stresses occurring in electrolytic solutions with non-homogenous concentrations arise from the joint effect of the electric force, due to the moving charges and the osmotic pressure. Using this theory as a basis, I succeeded in calculating the potential difference at the point of contact of two electrolytic solutions, and Nernst wrote to me later that my formula had been confirmed by his measurements.

In connection with the problems of the electric dissociation theory, I was soon also engaged in a voluminous correspondence with Wilhelm Ostwald, of Leipzig. Our correspondence led to many a critical debate, yet these were always carried on in the friendliest tone. Ostwald, by his very nature a firm believer in systematization, distinguished three different types of energy, corresponding to the three spatial dimensions, namely: Distance Energy, Surface Energy, and Space Energy. Distance Energy, according to him, was the force of gravitation; Surface Energy, the surface tension of liquids; and Space Energy, the volume energy. I replied, among other comments, that there was no such thing as a volume energy in the sense specified by Ostwald. For instance, the energy of an ideal gas does not in fact even depend on the volume, but on the temperature of the gas. If an ideal gas is made to expand without doing any work, its volume increases, but the energy remains unchanged, whereas according to Ostwald, its energy ought to decrease with the decrease of the pressure.

Another controversy arose with relation to the question of the analogy between the passage of heat from a higher to a lower temperature and the sinking of a weight from a greater to a smaller height. I had emphasized the need for a sharp distinction between these two processes, for they differed from each other as basically as did the First and Second Laws of Thermodynamics. However, this theory of mine was contradicted by the view universally accepted in those days. and I just could not make my fellow physicists see it my way. In fact, certain physicists actually regarded Clausius' reasoning as unnecessarily complicated and even confused and they refused, in particular, to admit the concept of irreversibility, and thereby to assign to heat a special position among the forms of energy. They created in opposition to Clausius' theory of thermodynamics, the so-called science of "Energetics." The first fundamental proposition of Energetics, exactly like that of Clausius' theory, expresses the principle of the conservation of energy; but its second proposition, which is supposed to formulate the direction of all occurrences, postulates a perfect analogy between the passing, of heat from a higher to a lower temperature and the sinking of a weight from a greater to a smaller height. A consequence of this point of view was that the assumption of irreversibility for proving the Second Law of Thermodynamics was declared to be unessential; furthermore, the existence of an absolute zero of temperature was disputed, on the ground that for temperature, just as for height, only differences can be measured.

It is one of the most painful experiences of my entire scientific life that I have but seldom – in fact, I might say, never – succeeded in gaining universal recognition for a new result, the truth of which I could demonstrate by a conclusive, albeit only theoretical proof. This is what happened this time, too. All my sound arguments fell on deaf ears. It was simply impossible to be heard against the authority of men like Ostwald, Helm and Mach. I was firmly convinced that my claim of the basic difference between the transmission of heat and the sinking of a weight would eventually be proved to be right. But the annoying thing was that I was not to have at all the satisfaction of seeing myself vindicated. The universal acceptance of my thesis was ultimately brought about by considerations of an altogether different sort, unrelated to the arguments which I had adduced in support of it – namely, by the atomic theory, as represented by Ludwig Boltzmann.

Boltzmann succeeded in establishing, for a given gas in a given state, a function, H, which has the property that its value constantly decreases with time. It suffices, therefore, to identify the negative value of this H with entropy, to arrive at the principle of the increase of entropy. This discovery demonstrates, at the same time, irreversibility to be a characteristic of the processes occurring in a gas.

As events transpired, therefore, my claim concerning the fundamental difference between heat conduction and a purely mechanical process was victorious over the view previously entertained by outstanding authorities. Nevertheless, my contribution to the struggle was entirely superfluous, for even without it the outcome would have been the same.

Obviously, this battle, in which Boltzmann and Ostwald represented the opposing views, was fought rather heatedly, and produced also many a drastic effect, for the two antagonists were each other's equals in quick repartee and natural wit. After all that I have related, in this duel of minds I could play only the part of a second to Boltzmann – a second whose services were evidently not appreciated, not even noticed, by him. For Boltzmann knew very well that my viewpoint was basically different from his. He was especially annoyed by the fact that I was not only indifferent but to a certain extent even hostile to the atomic theory which was the foundation of his entire research. The reason was that at that time, I regarded the principle of the increase of entropy as no less immutably valid than the principle of the conservation of energy itself, whereas Boltzmann treated the former merely as a law of probabilities – in other words, as a principle that could admit of exceptions. The value of function H might also increase at times. Boltzmann did not go into this point in the deduction of his "H-Theorem," and a talented pupil of mine, E. Zermelo, noted emphatically this gap in a strict proof of the theorem. In fact, Boltzmann omitted in his deduction every mention of the indispensable presupposition of the validity of his theorem – namely, the assumption of molecular disorder. He must have simply taken it for granted. At any rate, he answered young Zermelo in a tone of biting sarcasm, which was meant partly for me, too, for Zermelo's paper had been published with my approval. This was how Boltzmann assumed that ill-tempered tone which he continued to exhibit toward me, on later occasions as well, both in his publications and in our personal correspondence; and it was only in the last years of his life, when I informed him of the atomistic foundation for my radiation law, that he assumed a friendlier attitude.

Boltzmann eventually triumphed in the fight against Ostwald and the adherents of Energetics, as it had been self-evident to me that he would, in view of all that I have just mentioned. The basic difference between the conduction of heat and a purely mechanical process became universally recognized. This experience gave me also an opportunity to learn a fact – a remarkable one, in my opinion: A new scientific truth does not triumph, by convincing, its opponents and making them see the light, but rather because its opponents eventually die, and a new generation grows up that is familiar with it.

Otherwise, the controversies just mentioned held comparatively little interest for me, as they could not be expected to produce anything new. My attention, therefore, was soon claimed by quite another problem, which was to dominate me and urge me on to a great many different investigations for a long time to come. The measurements made by O. Lummer and E. Pringsheim in the German Physico-Technical Institute, in connection with the study of the thermal spectrum, directed my attention to Kirchhoff's Law, which says that in an evacuated cavity, bounded by totally reflecting walls, and containing any arbitrary number of emitting and absorbing bodies, in time a state will be reached where all bodies have the same temperature, and the radiation, in all its properties including its spectral energy distribution, depends not on the nature of the bodies, but solely and exclusively on the temperature. Thus, this so-called Normal Spectral Energy Distribution represents something absolute, and since I had always regarded the search for the absolute as the loftiest goal of all scientific activity, I eagerly set to work. I found a direct method for solving the problem in the application of Maxwell's Electromagnetic Theory of Light. Namely, I assumed the cavity to be filled with simple linear oscillators or resonators, subject to small damping forces and having different periods; and I expected the exchange of energy caused by the reciprocal radiation of the oscillators to result, in time, in a stationary state of the normal energy distribution corresponding to Kirchhoff's Law.

This extended series of investigations, certain ones of which could be verified by comparisons with known observational data, such as the measurements of damping by V. Bjerknes, resulted in establishing the general relationship between the energy of an oscillator having a definite period, and the energy radiation of the corresponding spectral region in the surrounding field when the exchange of energy is stationary. From this there followed the remarkable result that this relationship is absolutely independent of the damping constant of the oscillator – a circumstance which was very pleasing and welcome to me, because it permitted the entire problem to be simplified, by substituting the energy of the oscillator for the energy of the radiation, thus replacing a complicated structure possessing many degrees of freedom, by a simple system with just one degree of freedom.

To be sure, this result represented a mere preliminary to the tackling of the real problem, which now loomed all the more formidably before me. My first attempt to overcome it was unsuccessful, for my original silent hope that the radiation emitted by the oscillator would differ, in some characteristic way, from the absorbed radiation, turned out to have been mere wishful thinking. The oscillator reacts only to those rays which it is capable of emitting, and is completely insensitive to adjacent spectral regions.

Moreover, my suggestion that the oscillator was capable of exerting a unilateral, in other words irreversible, effect on the energy of the surrounding field, drew a vigorous protest from Boltzmann, who, with his wider experience in this domain, demonstrated that according to the laws of classical dynamics, each of the processes I considered could also take place in the opposite direction; and indeed in such a manner, that a spherical wave emitted by an oscillator could reverse its direction of motion, contract progressively until it reached the oscillator and be reabsorbed by the latter, so that the oscillator could then again emit the previously absorbed energy in the same direction from which the energy had been received. To be sure, I could exclude such odd phenomena as inwardly directed spherical waves, by the introduction of a specific stipulation – the hypothesis of a natural radiation, which plays the same part in the theory of radiation as the hypothesis of molecular disorder in the kinetic theory of gases, in that it guarantees the irreversibility of the radiation processes. But the calculations showed ever more clearly that an essential link was still missing, without which the attack on the core of the entire problem could not be undertaken successfully.

So I had no other alternative than to tackle the problem once again – this time from the opposite side, namely, from the side of thermodynamics, my own home territory where I felt myself to be on safer ground. In fact, my previous studies of the Second Law of Thermodynamics came to stand me in good stead now, for at the very outset I hit upon the idea of correlating not the temperature but the entropy of the oscillator with its energy. It was an odd jest of fate that a circumstance which on former occasions I had found unpleasant, namely, the lack of interest of my colleagues in the direction taken by my investigations, now turned out to be an outright boon. While a host of outstanding physicists worked on the problem of spectral energy distribution, both from the experimental and theoretical aspect, every one of them directed his efforts solely toward exhibiting the dependence of the intensity of radiation on the temperature. On the other hand, I suspected that the fundamental connection lies in the dependence of entropy upon energy. As the significance of the concept of entropy had not yet come to be fully appreciated, nobody paid any attention to the method adopted by me, and I could work out my calculations completely at my leisure, with absolute thoroughness, without fear of interference or competition.

Since for the irreversibility of the exchange of energy between an oscillator and the radiation activating it, the second differential quotient of its entropy with respect to its energy is of characteristic significance, I calculated the value of this function on the assumption that Wien's Law of the Spectral Energy Distribution is valid – a law which was then in the focus of general interest; I got the remarkable result that on this assumption the reciprocal of that value, which I shall call here R, is proportional to the energy. This relationship is so surprisingly simple that for a while I considered it to possess universal validity, and I endeavored to prove it theoretically. However, this view soon proved to be untenable in the face of later measurements. For although in the case of small energies and correspondingly short waves Wien's Law continued to be confirmed in a satisfactory manner, in the case of large values of the energy and correspondingly long waves, appreciable divergences were found, first by Lummer and Pringsheim; and finally the measurements of H. Rubens and F. Kurlbaum on infrared rays of fluorspar and rock-salt revealed a behavior which, though totally different, is again a simple one, in so far as the function R is proportional not to the energy but to the square of the energy for large values of the energy and wave-lengths.

Thus, direct experiments established two simple limits for the function R: For small energies, R is proportional to the energy; for larger energy values R is proportional to the square of the energy. Obviously, just as every principle of spectral energy distribution yields a certain value for R, so also every formula for R leads to a definite law of the distribution of energy. The problem was to find such a formula for R which, would result in the law of the distribution of energy established by measurements. Therefore, the most obvious step for the general case was to make the value of R equal to the sum of a term proportional to the first power of the energy and another term proportional to the second power of the energy, so that the first term becomes decisive for small values of the energy and the second term for large values. In this way a new radiation formula was obtained, and I submitted it for examination to the Berlin Physical Society, at the meeting on October 19, 1900.

The very next morning, I received a visit from my colleague Rubens. He came to tell me that after the conclusion of the meeting he had that very night checked my formula against the results of his measurements, and found a satisfactory concordance at every point. Also Lummer and Pringsheim, who first thought to have discovered divergences, soon withdrew their objections; for, as Pringsheim related it to me, the observed divergences turned out to have been due to an error in calculation. Later measurements, too, confirmed my radiation formula again and again, – the finer the methods of measurement used, the more accurate the formula was found to be.

But even if the absolutely precise validity of the radiation formula is taken for granted, so long as it had merely the standing of a law disclosed by a lucky intuition, it could not be expected to possess more than a formal significance. For this reason, on the very day when I formulated this law, I began to devote myself to the task of investing it with a true physical meaning. This quest automatically led me to study the interrelation of entropy and probability – in other words, to pursue the line of thought inaugurated by Boltzmann. Since the entropy S is an additive magnitude but the probability W is a multiplicative one, I simply postulated that $S = k \log W$, where k is a universal constant; and I investigated whether the formula for W, which is obtained when S is replaced by its value corresponding to the above radiation law, could be interpreted as a measure of probability.

As a result,[1] I found that this was actually possible, and

---

[1] As Max von Laue noted in a footnote in M. Planck, *Scientific Au-*

that in this connection k represents the so-called absolute
gas constant, referred not to gram-molecules or mols, but to
the real molecules. It is, understandably, often called Boltzmann's constant. However, this calls for the comment that
Boltzmann never introduced this constant, nor, to the best of
my knowledge, did he ever think of investigating its numerical value. For had he done so, he would have had to examine
the matter of the number of the real atoms – a task, however,
which he left to his colleague J. Loschmidt, while he, in his
own calculations, always kept in sight the possibility that the
kinetic theory of gases represents only a mechanical picture.
He was therefore satisfied with stopping at the gram-atoms.
The letter k has won acceptance only gradually. Even several
years after its introduction, it was still customary to calculate
with the Loschmidt number L.

Now as for the magnitude W, I found that in order to
interpret it as a probability, it was necessary to introduce
a universal constant, which I called h. Since it had the di-
mension of action (energy × time), I gave it the name, *elementary quantum of action*. Thus the nature of entropy as a
measure of probability, in the sense indicated by Boltzmann,
was established in the domain of radiation, too. This was
made especially clear in a proposition, the validity of which
my closest pupil, Max von Laue, convinced me in a number
of conversations – namely, that the entropy of two coherent
pencils of light is smaller than the sum of the entropies of the
individual pencils of rays, quite consistently with the proposition that the probability of the happening of two mutually
interdependent reactions is different from the product of the
individual reactions.

tobiography And Other Papers, translated by F. Gaynor (Williams &
Norgate Ltd, London 1950), p. 42: "This finding, containing the introduction of the ultimate energy quanta for the oscillator, was reported
by Max Planck again before the Physical Society of Berlin on December
14, 1900. That was the birthday of the Quantum Theory."

While the significance of the quantum of action for the interrelation between entropy and probability was thus conclusively established, the part played by this new constant in the uniformly regular occurrence of physical processes still remained an open question. I therefore, tried immediately to weld the elementary quantum of action h somehow into the framework of the classical theory. But in the face of all such attempts, this constant showed itself to be obdurate. So long as it could be regarded as infinitesimally small, i.e. when dealing with higher energies and longer periods of time, everything was in perfect order. But in the general case difficulties would arise at one point or another, difficulties, which became more noticeable as higher frequencies were taken into consideration, The failure of every attempt to bridge this obstacle soon made it evident that the elementary quantum of action plays a fundamental part in atomic physics, and that its introduction opened up a new era in natural science. For it heralded the advent of something entirely unprecedented, and was destined to remodel basically the physical outlook and thinking of man which, ever since Leibniz and Newton laid the groundwork for infinitesimal calculus, were founded on the assumption that all causal interactions are continuous.

My futile attempts to fit the elementary quantum of action somehow into the classical theory continued for a number of years, and they cost me a great deal of effort. Many of my colleagues saw in this something bordering on a tragedy. But I feel differently about it. For the thorough enlightenment I thus received was all the more valuable. I now knew for a fact that the elementary quantum of action played a far more significant part in physics than I had originally been inclined to suspect, and this recognition made me see clearly the need for the introduction of totally new methods of analysis and reasoning in the treatment of atomic problems. The development of such methods – in which, however, I could no longer take an active part – was advanced mainly by the efforts of Niels

Bohr and Erwin Schrödinger. Bohr, with his atom model and Correspondence Principle, laid the foundation for a reasonable unification of quantum theory with classical theory. Schrödinger, through his differential equation, created wave mechanics, and thereby the dualism between wave and particle.

I have just described how the quantum theory came gradually to occupy the focus of my interest in the field of physics. Eventually, it had to share this prominent position with another principle, which introduced me to a new sphere of ideas. In 1905, Albert Einstein published a paper in the *Annalen der Physik* which contained the basic ideas of the Theory of Relativity, and my acute interest in their development was immediately roused.

In order to preclude a likely misunderstanding, I have to insert here a few explanatory remarks of general character. In the opening paragraph of this autobiographical sketch, I emphasized that I had always looked upon the search for the absolute as the noblest and most worth while task of science. The reader might consider this contradictory to my avowed interest in the Theory of Relativity. But it would be fundamentally erroneous to look at it that way. For everything that is relative presupposes the existence of something that is absolute, and is meaningful only when juxtaposed to something absolute. The oftenheard phrase, "Everything is relative," is both misleading and thoughtless. The Theory of Relativity, too, is based on something absolute, namely, the determination of the matrix of the space-time continuum; and it is an especially stimulating undertaking to discover the absolute which alone makes meaningful something given as relative.

Our every starting-point must necessarily be something relative. All our measurements are relative. The material that goes into our instruments varies according to its geographic source; their construction depends on the skill of the designer and toolmaker; their manipulation is contingent on the special purposes pursued by the experimenter. Our task is to find in all these factors and data, the absolute, the universally valid, the invariant, that is hidden in them.

This applies to the Theory of Relativity, too. I was attracted by the problem of deducing from its propositions that which served as their absolute immutable foundation. The way in which this was accomplished, was comparatively simple. In the first place, the Theory of Relativity confers an absolute meaning on a magnitude which in classical theory has only a relative significance: the velocity of light. The velocity of light is to the Theory of Relativity as the elementary quantum of action is to the Quantum Theory: it is its absolute core. In this connection, it turns out that a general principle of classical theory, the least-action principle, is also invariant with respect to the Theory of Relativity; accordingly, the quantum of action retains its significance in the Theory of Relativity as well.

This was what I tried to establish in all details, first for point masses, and then for black-body radiation. These researches yielded, among other results, the inertia of radiation and the invariance of entropy in systems possessing relative velocities.

But this is not all. The absolute showed itself to be even more deeply rooted in the order of natural laws than had been assumed for a long time. In 1906, W. Nernst came out with his heat theorem, often referred to as the "Third Law of Thermodynamics." As I immediately established, it amounted to the hypothesis that entropy, until then defined only up to an additive constant, possessed an absolute positive value. This value, from which all equations of equilibrium follow, can be calculated beforehand. In the case of a chemically homogeneous solid or liquid (in other words, a solid or liquid composed of homogeneous molecules) of which the absolute temperature is zero, this value is likewise zero. This principle in itself expresses an important fact, namely that the specific heat of a solid or liquid vanishes at the absolute zero of temperature. For other temperatures, fruitful inferences follow, with respect to the melting points of a body and the transition temperature of allotropic changes. If now one passes from chemically homogeneous solids and liquids to bodies with heterogeneous molecules, or to solutions and gases, the absolute entropy is calculated by means of combinatory considerations, in which the elementary quantum of action, too, must be included. In this way one can obtain the chemical properties of any given body, and thus a complete answer is found to all problems dealing with physico-chemical equilibrium. However, in questions concerned with the temporal developments of processes other forces must be taken into account, and problems about such forces are not resolved by considering the value of the entropy,

Even though as a consequence of my advancing age I have been able to take an increasingly smaller direct part in scientific research, there was compensation for this in the considerable expansion of my scientific correspondence, which I found enormously stimulating and invigorating. In this respect, I would like to mention, in particular, my correspondence with Cl. Schaefer, whose *Introduction to Theoretical Physics* I consider as pedagogically unexcelled. Our correspondence concerned his presentation of the Second Law of Thermodynamics. I also carried on an interesting correspondence with A. Sommerfeld, on the problem of the quantization of systems with several degrees of freedom. This particular correspondence even culminated in a final exchange of poetic tributes, which I shall take the liberty to quote here, although I must demur in all fairness that Sommerfeld seriously underestimated his own achievements in this field. This is how he referred to my studies on the structures of phase space:

> *You cultivate the virgin soil,*
> *Where picking flowers was my only toil.*

My only possible reply was:

> *You picked flowers — well, so have I.*
> *Let them be, then, combined;*
> *Let us exchange our flowers fair.*
> *And in the brightest wreath them bind.*

I have satisfied my inner need for bearing witness, as fully as possible, both to the results of my scientific labors and to my gradually crystallized attitude to general questions – such as the meaning of exact science, its relationship to religion, the connection between causality and free will – by always complying willingly with the ever increasing number of invitations to deliver lectures before Academies, Universities, learned societies, and before the general public, and these lectures have been the source of a many a valuable personal stimulation which I shall gratefully cherish in loving memory for the rest of my life.

Part II

Eight Lectures on
Theoretical Physics

PREFACE TO THE ORIGINAL EDITION

The present book has for its object the presentation of the lectures which I delivered as foreign lecturer at Columbia University in the spring of the present year under the title: "The Present System of Theoretical Physics." The points of view which influenced me in the selection and treatment of the material are given at the beginning of the first lecture. Essentially, they represent the extension of a theoretical physical scheme, the fundamental elements of which I developed in an address at Leyden entitled: "The Unity of the Physical Concept of the Universe." Therefore I regard it as advantageous to consider again some of the topics of that lecture. The presentation will not and can not, of course, claim to cover exhaustively in all directions the principles of theoretical physics.

Berlin, 1909   The Author

30

1 INTRODUCTION: REVERSIBILITY AND IRREVERSIBILITY

Colleagues, ladies and gentlemen: The cordial invitation, which the President of Columbia University extended to me to deliver at this prominent center of American science some lectures in the domain of theoretical physics, has inspired in me a sense of the high honor and distinction thus conferred upon me and, in no less degree, a consciousness of the special obligations which, through its acceptance, would be imposed upon me. If I am to count upon meeting in some measure your just expectations, I can succeed only through directing your attention to the branches of my science with which I myself have been specially and deeply concerned, thus exposing myself to the danger that my report in certain respects shall thereby have somewhat too subjective a coloring.

From those points of view which appear to me the most striking, it is my desire to depict for you in these lectures the present status of the system of theoretical physics. I do not say: the present status of theoretical physics; for to cover this far broader subject, even approximately, the number of lecture hours at my disposal would by no means suffice. Time limitations forbid the extensive consideration of the details of this great field of learning; but it will be quite possible to develop for you, in bold outline, a representation of the system as a whole, that is, to give a sketch of the fundamental laws which rule in the physics of today, of the most important hypotheses employed, and of the great ideas which have recently forced themselves into the subject. I will often gladly endeavor to go into details, but not in the sense of a thorough treatment of the subject, and only with the object of making the general laws more clear, through appropriate specially chosen examples. I shall select these examples from the most varied branches of physics.

If we wish to obtain a correct understanding of the achievements of theoretical physics, we must guard in equal measure against the mistake of overestimating these achievements, and on the other hand, against the corresponding mistake of underestimating them. That the second mistake is actually often made, is shown by the circumstance that quite recently voices have been loudly raised maintaining the bankruptcy and, débacle of the whole of natural science. But I think such assertions may easily be refuted by reference to the simple fact that with each decade the number and the significance of the means increase, whereby mankind learns directly through the aid of theoretical physics to make nature useful for its own purposes. The technology of today would be impossible without the aid of theoretical physics. The development of the whole of electrotechnics from galvanoplasty to wireless telegraphy is a striking proof of this, not to mention aerial navigation. On the other hand, the mistake of overestimating the achievements of theoretical physics appears to me to be much more dangerous, and this danger is particularly threatened by those who have penetrated comparatively little into the heart of the subject. They maintain that some time, through a proper improvement of our science, it will be possible, not only to represent completely through physical formulae the inner constitution of the atoms, but also the laws of mental life. I think that there is nothing in the world entitling us to the one or the other of these expectations. On the other hand, I believe that there is much which directly opposes them. Let us endeavor then to follow the middle course and not to deviate appreciably toward the one side or the other.

When we seek for a solid immovable foundation which is able to carry the whole structure of theoretical physics, we meet with the questions: What lies at the bottom of physics? What is the material with which it operates? Fortunately, there is a complete answer to this question. The material with which theoretical physics operates is measurements, and mathematics is the chief tool with which this material is worked. All physical ideas depend upon measurements, more or less exactly carried out, and each physical definition, each physical law, possesses a more definite significance the nearer it can be brought into accord with the results of measurements. Now measurements are made with the aid of the senses; before all with that of sight, with hearing and with feeling. Thus far, one can say that the origin and the foundation of all physical research are seated in our sense perceptions. Through sense perceptions only do we experience anything of nature; they are the highest court of appeal in questions under dispute. This view is completely confirmed by a glance at the historical development of physical science. Physics grows upon the ground of sensations. The first physical ideas derived were from the individual perceptions of man, and, accordingly, physics was subdivided into: physics of the eye (optics), physics of the ear (acoustics), and physics of heat sensation (theory of heat). It may well be said that so far as there was a domain of sense, so far extended originally the domain of physics. Therefore it appears that in the beginning the division of physics was based upon the peculiarities of man. It possessed, in short, an anthropomorphic character. This appears also, in that physical research, when not occupied with special sense perceptions, is concerned with practical life, and particularly with the practical needs of men. Thus, the art of geodesy led to geometry,

the study of machinery to mechanics, and the conclusion lies
near that physics in the last analysis had only to do with the
sense perceptions and needs of mankind.

In accordance with this view, the sense perceptions are
the essential elements of the world; to construct an object
as opposed to sense perceptions is more or less an arbitrary
matter of will. In fact, when I speak of a tree, I really mean
only a complex of sense perceptions: I can see it, I can hear
the rustling of its branches, I can smell its fragrance, I experience pain if I knock my head against it, but disregarding
all of these sensations, there remains nothing to be made the
object of a measurement, wherewith, therefore, natural science can occupy itself. This is certainly true. In accordance
with this view, the problem of physics consists only in the relating of sense perceptions, in accordance with experience, to
fixed laws; or, as one may express it, in the greatest possible
economic accommodation of our ideas to our sensations, an
operation which we undertake solely because it is of use to us
in the general battle of existence.

All this appears extraordinarily simple and clear and, in
accordance with it, the fact may readily be explained that
this positivist view is quite widely spread in scientific circles
today. It permits, so far as it is limited to the standpoint
here depicted (not always done consistently by the exponents
of positivism), no hypothesis, no metaphysics; all is clear and
plain. I will go still further; this conception never leads to
an actual contradiction. I may even say, it can lead to no
contradiction. But, ladies and gentlemen, this view has never
contributed to any advance in physics. If physics is to advance, in a certain sense its problem must be stated in quite
the inverse way, on account of the fact that this conception is
inadequate and at bottom possesses only a formal meaning.

The proof of the correctness of this assertion is to be found
directly from a consideration of the process of development
which theoretical physics has actually undergone, and which

one certainly cannot fail to designate as essential. Let us compare the system of physics of today with the earlier and more primitive system which I have depicted above. At the first glance we encounter the most striking difference of all, that in the present system, as well in the division of the various physical domains as in all physical definitions, the historical element plays a much smaller role than in the earlier system. While originally, as I have shown above, the fundamental ideas of physics were taken from the specific sense perceptions of man, the latter are today in large measure excluded from physical acoustics, optics, and the theory of heat. The physical definitions of tone, color, and of temperature are today in no wise derived from perception through the corresponding senses; but tone and color are defined through a vibration number or wave length, and the temperature through the volume change of a thermometric substance, or through a temperature scale based on the second law of thermodynamics; but heat sensation is in no wise mentioned in connection with the temperature. With the idea of force it has not been otherwise. Without doubt, the word force originally meant bodily force, corresponding to the circumstance that the oldest tools, the ax, hammer, and mallet, were swung by man's hands, and that the first machines, the lever, roller, and screw, were operated by men or animals. This shows that the idea of force was originally derived from the sense of force, or muscular sense, and was, therefore, a specific sense perception. Consequently, I regard it today as quite essential in a lecture on mechanics to refer, at any rate in the introduction, to the original meaning of the force idea. But in the modern exact definition of force the specific notion of sense perception is eliminated, as in the case of color sense, and we may say, quite in general, that in modern theoretical physics the specific sense perceptions play a much smaller role in all physical definitions than formerly. In fact, the crowding into the background of the specific sense elements goes so far that the branches of physics which were originally completely and uniquely characterized by an arrangement in accordance with definite sense perceptions have fallen apart, in consequence of the loosening of the bonds between different and widely separated branches, on account of the general advance towards simplification and coordination. The best example of this is furnished by the theory of heat. Earlier, heat formed a separate and unified domain of physics, characterized through the perceptions of heat sensation. Today one finds in well nigh all physics textbooks dealing with heat a whole domain, that of radiant heat, separated and treated under optics. The significance of heat perception no longer suffices to bring together the heterogeneous parts.

In short, we may say that the characteristic feature of the entire previous development of theoretical physics is a definite elimination from all physical ideas of the anthropomorphic elements, particularly those of specific sense perceptions. On the other hand, as we have seen above, if one reflects that the perceptions form the point of departure in all physical research, and that it is impossible to contemplate their absolute exclusion, because we cannot close the source of all our knowledge, then this conscious departure from the original conceptions must always appear astonishing or even paradoxical. There is scarcely a fact in the history of physics which today stands out so clearly as this. Now, what are the great advantages to be gained through such a real obliteration of personality? What is the result for the sake of whose achievement are sacrificed the directness and succinctness such as only the special sense perceptions vouchsafe to physical ideas?

The result is nothing more than the attainment of unity and compactness in our system of theoretical physics, and, in fact, the unity of the system, not only in relation to all of its details, but also in relation to physicists of all places, all times, all peoples, all cultures. Certainly, the system of theoretical physics should be adequate, not only for the inhabitants of this earth, but also for the inhabitants of other heavenly bodies. Whether the inhabitants of Mars, in case such actually exist, have eyes and ears like our own, we do not know,—it is quite improbable; but that they, in so far as they possess the necessary intelligence, recognize the law of gravitation and the principle of energy, most physicists would hold as self evident: and anyone to whom this is not evident had better not appeal to the physicists, for it will always remain for him an unsolvable riddle that the same physics is made in the United States as in Germany.

To sum up, we may say that the characteristic feature of the actual development of the system of theoretical physics is an ever extending emancipation from the anthropomorphic elements, which has for its object the most complete separation possible of the system of physics and the individual personality of the physicist. One may call this the objectiveness of the system of physics. In order to exclude the possibility of any misunderstanding, I wish to emphasize particularly that we have here to do, not with an absolute separation of physics from the physicist—for a physics without the physicist is unthinkable,—but with the elimination of the individuality of the particular physicist and therefore with the production of a common system of physics for all physicists.

Now, how does this principle agree with the positivist conceptions mentioned above? Separation of the system of physics from the individual personality of the physicist? Opposed to this principle, in accordance with those conceptions, each particular physicist must have his special system of physics, in case that complete elimination of all metaphysical elements is effected; for physics occupies itself only with the facts discovered through perceptions, and only the individual perceptions are directly involved. That other living beings have sensations is, strictly speaking, but a very probable, though arbitrary, conclusion from analogy. The system of physics is therefore primarily an individual matter and, if two physicists accept the same system, it is a very happy circumstance in connection with their personal relationship, but it is not essentially necessary. One can regard this view-point however he will; in physics it is certainly quite fruitless, and this is all that I care to maintain here. Certainly, I might add, each great physical idea means a further advance toward the emancipation from anthropomorphic ideas. This was true in the passage from the Ptolemaic to the Copernican cosmical system, just as it is true at the present time for the apparently impending passage from the so-called classical mechanics of mass points to the general dynamics originating in the principle of relativity. In accordance with this, man and the earth upon which he dwells are removed from the centre of the world. It may be predicted that in this century the idea of time will be divested of the absolute character with which men have been accustomed to endow it (cf. the final lecture). Certainly, the sacrifices demanded by every such revolution in the intuitive point of view are enormous; consequently, the resistance against such a change is very great. But the development of science is not to be permanently halted thereby; on the contrary, its strongest impetus is experienced through precisely those forces which attain success in the struggle against the old points of view, and to this extent such a struggle is constantly necessary and useful.

Now, how far have we advanced today toward the unification of our system of physics? The numerous independent domains of the earlier physics now appear reduced to two; mechanics and electrodynamics, or, as one may say: the physics of material bodies and the physics of the ether. The former comprehends acoustics, phenomena in material bodies, and chemical phenomena; the latter, magnetism, optics, and radiant heat. But is this division a fundamental one? Will it prove final? This is a question of great consequence for the future development of physics. For myself, I believe it must be answered in the negative, and upon the following grounds: mechanics and electrodynamics cannot be permanently sharply differentiated from each other. Does the process of light emission, for example, belong to mechanics or to electrodynamics? To which domain shall be assigned the laws of motion of electrons? At first glance, one may perhaps say: to electrodynamics, since with the electrons ponderable matter does not play any role. But let one direct his attention to the motion of free electrons in metals. There he will find, in the study of the classical researches of H. A. Lorentz, for example, that the laws obeyed by the electrons belong rather to the kinetic theory of gases than to electrodynamics. In general, it appears to me that the original differences between processes in the ether and processes in material bodies are to be considered as disappearing. Electrodynamics and mechanics are not so remarkably far apart, as is considered to be the case by many people, who already speak of a conflict between the mechanical and the electrodynamic views of the world. Mechanics requires for its foundation essentially nothing more than the ideas of space, of time, and of that which is moving, whether one considers this as a substance or a state. The same ideas are also involved in electrodynamics. A sufficiently generalized conception of mechanics can therefore also well include electrodynamics, and, in fact, there are many indications pointing toward the ultimate amalgamation of these two subjects, the domains of which already overlap in some measure.

If, therefore, the gulf between ether and matter be once bridged, what is the point of view which in the last analysis will best serve in the subdivision of the system of physics? The answer to this question will characterize the whole nature of the further development of our science. It is, therefore, the most important among all those which I propose to treat today. But for the purposes of a closer investigation it is necessary that we go somewhat more deeply into the peculiarities of physical principles.

We shall best begin at that point from which the first step was made toward the actual realization of the unified system of physics previously postulated by the philosophers only; at the principle of conservation of energy. For the idea of energy is the only one besides those of space and time which is common to all the various domains of physics. In accordance with what I have stated above, it will be apparent and quite self evident to you that the principle of energy, before its general formularization by Mayer, Joule, and Helmholz, also bore an anthropomorphic character. The roots of this principle lay already in the recognition of the fact that no one is able to obtain useful work from nothing; and this recognition had originated essentially in the experiences which were gathered in attempts at the solution of a technical problem: the discovery of perpetual motion. To this extent, perpetual motion has come to have for physics a far reaching significance, similar to that of alchemy for the chemist, although it was not the positive, but rather the negative results of these experiments, through which science was advanced. Today we speak of the principle of energy quite without reference to the technical viewpoint or to that of man. We say that the total amount of energy of an isolated system of bodies is a quantity whose amount can be neither increased nor diminished through any kind of process within the system, and we no longer consider the accuracy with which this law holds as dependent upon the refinement of the methods, which we at present possess, of testing experimentally the question of the realization of perpetual motion. In this, strictly speaking, unprovable generalization, impressed upon us with elemental force, lies the emancipation from the anthropomorphic elements mentioned above.

While the principle of energy stands before us as a complete independent structure, freed from and independent of the accidents appertaining to its historical development, this is by no means true in equal measure in the case of that principle which R. Clausius introduced into physics; namely, the second law of thermodynamics. This law plays a very peculiar role in the development of physical science, to the extent that one is not able to assert today that for it a generally recognized, and therefore objective formularization, has been found. In our present consideration it is therefore a matter of particular interest to examine more closely its significance.

In contrast to the first law of thermodynamics, or the energy principle, the second law may be characterized as follows. While the first law permits in all processes of nature neither the creation nor destruction of energy, but permits of transformations only, the second law goes still further into the limitation of the possible processes of nature, in that it permits, not all kinds of transformations, but only certain types, subject to certain conditions. The second law occupies itself, therefore, with the question of the kind and, in particular, with the direction of any natural process.

At this point a mistake has frequently been made, which has hindered in a very pronounced manner the advance of science up to the present day. In the endeavor to give to the second law of thermodynamics the most general character possible, it has been proclaimed by followers of W. Ostwald as the second law of energetics, and the attempt made so to formulate it that it shall determine quite generally the direction of every process occurring in nature. Some weeks ago I read in a public academic address of an esteemed colleague the statement that the import of the second law consists in this, that a stone falls downwards, that water flows not up hill, but down, that electricity flows from a higher to a lower potential, and so on. This is a mistake which at present is altogether too prevalent not to warrant mention here.

The truth is, these statements are false. A stone can just as well rise in the air as fall downwards; water can likewise flow upwards, as, for example, in a spring; electricity can flow very well from a lower to a higher potential, as in the case of oscillating discharge of a condenser. The statements are obviously quite correct, if one applies them to a stone originally at rest, to water at rest, to electricity at rest; but then they follow immediately from the energy principle, and one does not need to add a special second law. For, in accordance with the energy principle, the kinetic energy of the stone or of the water can only originate at the cost of gravitational energy, i.e., the center of mass must descend. If, therefore, motion is to take place at all, it is necessary that the gravitational energy shall decrease. That is, the center of mass must descend. In like manner, an electric current between two condenser plates can originate only at the cost of electrical energy already present; the electricity must therefore pass to a lower potential. If, however, motion and current be already present, then one is not able to say, a priori, anything in regard to the direction of the change; it can take place just as well in one direction as the other. Therefore, there is no new insight into nature to be obtained from this point of view.

Upon an equally inadequate basis rests another conception of the second law, which I shall now mention. In considering the circumstance that mechanical work may very easily be transformed into heat, as by friction, while on the other hand heat can only with difficulty be transformed into work, the attempt has been made so to characterize the second law, that in nature the transformation of work into heat can take place completely, while that of heat into work, on the other hand, only incompletely and in such manner that every time a quantity of heat is transformed into work another corresponding quantity of energy must necessarily undergo at the same time a compensating transformation, as, e.g., the passage of heat from a higher to a lower temperature. This assertion is in certain special cases correct, but does not strike in general at the true import of the matter, as I shall show by a simple example.

One of the most important laws of thermodynamics is,

an ideal gas depends only upon its
ḻpon its volume. If an ideal gas be al-
ᴸle doing work, and if the cooling of the
through the simultaneous addition of heat
ᴸervoir at higher temperature, the gas remains
ᴸ temperature and energy content, and one may
ᴸe heat furnished by the heat reservoir is completely
ᴸmed into work without exchange of energy. Not the
ᴸbjection can be urged against this assertion. The law of
ᴸmplete transformation of heat into work is retained only
ᴸhrough the adoption of a different point of view, but which
has nothing to do with the status of the physical facts and
only modifies the way of looking at the matter, and therefore can neither be supported nor contradicted through facts;
namely, through the introduction ad hoc of new particular
kinds of energy, in that one divides the energy of the gas into
numerous parts which individually can depend upon the volume. But it is a priori evident that one can never derive from
so artificial a definition a new physical law, and it is with
such that we have to do when we pass from the first law, the
principle of conservation of energy, to the second law.

I desire now to introduce such a new physical law: "It
is not possible to construct a periodically functioning motor
which in principle does not involve more than the raising of
a load and the cooling of a heat reservoir." It is to be understood, that in one cycle of the motor quite arbitrary complicated processes may take place, but that after the completion
of one cycle there shall remain no other changes in the surroundings than that the heat reservoir is cooled and that the
load is raised a corresponding distance, which may be calculated from the first law. Such a motor could of course be
used at the same time as a refrigerating machine also, without any further expenditure of energy and materials. Such
a motor would moreover be the most efficient in the world,
since it would involve no cost to run it; for the earth, the

atmosphere, or the ocean could be utilized as the heat r
voir. We shall call this, in accordance with the proposa
W. Ostwald, perpetual motion of the second kind. Whether
nature such a motion is actually possible cannot be inferre
from the energy principle, and may only be determined by
special experiments.

Just as the impossibility of perpetual motion of the first
kind leads to the principle of the conservation of energy, the
quite independent principle of the impossibility of perpetual
motion of the second kind leads to the second law of thermodynamics, and, if we assume this impossibility as proven
experimentally, the general law follows immediately: *there
are processes in nature which in no possible way can be made
completely reversible.* For consider, e.g., a frictional process
through which mechanical work is transformed into heat with
the aid of suitable apparatus, if it were actually possible to
make in some way such complicated apparatus completely
reversible, so that everywhere in nature exactly the same
conditions be reestablished as existed at the beginning of
the frictional process, then the apparatus considered would
be nothing more than the motor described above, furnishing
a perpetual motion of the second kind. This appears evident immediately, if one clearly perceives what the apparatus
would accomplish: transformation of heat into work without
any further outstanding change.

We call such a process, which in no wise can be made
completely reversible, an irreversible process, and all other
processes reversible processes; and thus we strike the kernel
of the second law of thermodynamics when we say that irreversible processes occur in nature. In accordance with this,
the changes in nature have a unidirectional tendency. With
each irreversible process the world takes a step forward, the
traces of which under no circumstances can be completely
obliterated. Besides friction, examples of irreversible processes are: heat conduction, diffusion, conduction of electric-

ity in conductors of finite resistance, emission of light and heat radiation, disintegration of the atom in radioactive substances, and so on. On the other hand, examples of reversible processes are: motion of the planets, free fall in empty space, the undamped motion of a pendulum, the frictionless flow of liquids, the propagation of light and sound waves without absorption and refraction, undamped electrical vibrations, and so on. For all these processes are already periodic or may be made completely reversible through suitable contrivances, so that there remains no outstanding change in nature; for example, the free fall of a body whereby the acquired velocity is utilized to raise the body again to its original height; a light or sound wave which is allowed in a suitable manner to be totally reflected from a perfect mirror.

What now are the general properties and criteria of irreversible processes, and what is the general quantitative measure of irreversibility? This question has been examined and answered in the most widely different ways, and it is evident here again how difficult it is to reach a correct formularization of a problem. Just as originally we came upon the trail of the energy principle through the technical problem of perpetual motion, so again a technical problem, namely, that of the steam engine, led to the differentiation between reversible and irreversible processes. Long ago Sadi Carnot recognized, although he utilized an incorrect conception of the nature of heat, that irreversible processes are less economical than reversible, or that in an irreversible process a certain opportunity to derive mechanical work from heat is lost. What then could have been simpler than the thought of making, quite in general, the measure of the irreversibility of a process the quantity of mechanical work which is unavoidably lost in the process. For a reversible process then, the unavoidably lost work is naturally to be set equal to zero. This view, in accordance with which the import of the second law consists in a dissipation of useful energy, has in fact, in certain special cases, e.g., in isothermal processes, proved itself useful. It has persisted, therefore, in certain of its aspects up to the present day; but for the general case, however, it has shown itself as fruitless and, in fact, misleading. The reason for this lies in the fact that the question concerning the lost work in a given irreversible process is by no means to be answered in a determinate manner, so long as nothing further is specified with regard to the source of energy from which the work considered shall be obtained.

An example will make this clear. Heat conduction is an irreversible process, or as Clausius expresses it: Heat cannot without compensation pass from a colder to a warmer body. What now is the work which in accordance with definition is lost when the quantity of heat Q passes through direct conduction from a warmer body at the temperature T_1 to a colder body, at the temperature T_2? In order to answer this question, we make use of the heat transfer involved in carrying out a reversible Carnot cyclical process between the two bodies employed as heat reservoirs. In this process a certain amount of work would be obtained, and it is just the amount sought, since it is that which would be lost in the direct passage by conduction; but this has no definite value so long as we do not know whence the work originates, whether, e.g., in the warmer body or in the colder body, or from somewhere else. Let one reflect that the heat given up by the warmer body in the reversible process is certainly not equal to the heat absorbed by the colder body, because a certain amount of heat is transformed into work, and that we can identify, with exactly the same right, the quantity of heat Q transferred by the direct process of conduction with that which in the cyclical process is given up by the warmer body, or with that absorbed by the colder body. As one does the former or the latter, he accordingly obtains for the quantity of lost work in the process of conduction:

$$Q\frac{T_1 - T_2}{T_1} \quad \text{or} \quad Q\frac{T_1 - T_2}{T_2}.$$

We see, therefore, that the proposed method of expressing mathematically the irreversibility of a process does not in general effect its object, and at the same time we recognize the peculiar reason which prevents its doing so. The statement of the question is too anthropomorphic. It is primarily too much concerned with the needs of mankind, in that it refers directly to the acquirement of useful work. If one require from nature a determinate answer, he must take a more general point of view, more disinterested, less economic. We shall now seek to do this.

Let us consider any typical process occurring in nature. This will carry all bodies concerned in it from a determinate initial state, which I designate as state A, into a determinate final state B. The process is either reversible or irreversible. A third possibility is excluded. But whether it is reversible or irreversible depends solely upon the nature of the two states A and B, and not at all upon the way in which the process has been carried out; for we are only concerned with the answer to the question as to whether or not, when the state B is once reached, a complete return to A in any conceivable manner may be accomplished. If now, the complete return from B to A is not possible, and the process therefore irreversible, it is obvious that the state B may be distinguished in nature through a certain property from state A. Several years ago I ventured to express this as follows: that nature possesses a greater "preference" for state B than for state A. In accordance with this mode of expression, all those processes of nature are impossible for whose final state nature possesses a smaller preference than for the original state. Reversible processes constitute a limiting case; for such, nature possesses an equal preference for the initial and for the final state, and the passage between them takes place as well in one direction as the other.

We have now to seek a physical quantity whose magnitude shall serve as a general measure of the preference of nature for a given state. This quantity must be one which is directly determined by the state of the system considered, without reference to the previous history of the system, as is the case with the energy, with the volume, and with other properties of the system. It should possess the peculiarity of increasing in all irreversible processes and of remaining unchanged in all reversible processes, and the amount of change which it experiences in a process would furnish a general measure for the irreversibility of the process.

R. Clausius actually found this quantity and called it "entropy." Every system of bodies possesses in each of its states a definite entropy, and this entropy expresses the preference of nature for the state in question. It can, in all the processes which take place within the system, only increase and never decrease. If it be desired to consider a process in which external actions upon the system are present, it is necessary to consider those bodies in which these actions originate as constituting part of the system; then the law as stated in the above form is valid. In accordance with it, the entropy of a system of bodies is simply equal to the sum of the entropies of the individual bodies, and the entropy of a single body is, in accordance with Clausius, found by the aid of a certain reversible process. Conduction of heat to a body increases its entropy, and, in fact, by an amount equal to the ratio of the quantity of heat given the body to its temperature. Simple compression, on the other hand, does not change the entropy.

Returning to the example mentioned above, in which the quantity of heat Q is conducted from a warmer body at the temperature T_1 to a colder body at the temperature T_2, in accordance with what precedes, the entropy of the warmer body decreases in this process, while, on the other hand, that of the colder increases, and the sum of both changes, that is, the change of the total entropy of both bodies, is:

$$-\frac{Q}{T_1} + \frac{Q}{T_2} > 0.$$

This positive quantity furnishes, in a manner free from all arbitrary assumptions, the measure of the irreversibility of the process of heat conduction. Such examples may be cited indefinitely. Every chemical process furnishes an increase of entropy.

We shall here consider only the most general case treated by Clausius: an arbitrary reversible or irreversible cyclical process, carried out with any physico-chemical arrangement, utilizing an arbitrary number of heat reservoirs. Since the arrangement at the conclusion of the cyclical process is the same as that at the beginning, the final state of the process is to be distinguished from the initial state solely through the different heat content of the heat reservoirs, and in that a certain amount of mechanical work has been furnished or consumed. Let Q be the heat given up in the course of the process by a heat reservoir at the temperature T, and let A be the total work yielded (consisting, e.g., in the raising of weights); then, in accordance with the first law of thermodynamics:

$$\sum Q = A.$$

In accordance with the second law, the sum of the changes in entropy of all the heat reservoirs is positive, or zero. It follows, therefore, since the entropy of a reservoir is decreased by the amount Q/T through the loss of heat Q that:

$$\sum \frac{Q}{T} \leqq 0.$$

This is the well-known inequality of Clausius.

In an isothermal cyclical process, T is the same for all reservoirs. Therefore:

$$\sum Q \leqq 0, \quad \text{hence:} \quad A \leqq 0.$$

That is: in an isothermal cyclical process, heat is produced and work is consumed. In the limiting case, a reversible isothermal cyclical process, the sign of equality holds, and therefore the work consumed is zero, and also the heat produced. This law plays a leading role in the application of thermodynamics to physical chemistry.

The second law of thermodynamics including all of its consequences has thus led to the principle of increase of entropy. You will now readily understand, having regard to the questions mentioned above, why I express it as my opinion that in the theoretical physics of the future the first and most important differentiation of all physical processes will be into reversible and irreversible processes.

In fact, all reversible processes, whether they take place in material bodies, in the ether, or in both together, show a much greater similarity among themselves than to any irreversible process. In the differential equations of reversible processes the time differential enters only as an even power, corresponding to the circumstance that the sign of time can be reversed. This holds equally well for vibrations of the pendulum, electrical vibrations, acoustic and optical waves, and for motions of mass points or of electrons, if we only exclude every kind of damping. But to such processes also belong those infinitely slow processes of thermodynamics which consist of states of equilibrium in which the time in general plays no role, or, as one may also say, occurs with the zero power, which is to be reckoned as an even power. As Helmholtz has pointed out, all these reversible processes have the common property that they may be completely represented by the principle of least action, which gives a definite answer to all questions concerning any such measurable process, and, to this extent, the theory of reversible processes may be regarded as completely established. Reversible processes have, however, the disadvantage that singly and collectively they are only ideal: in actual nature there is no such thing as a reversible process. Every natural process involves in greater or less degree friction or conduction of heat. But in the domain of irreversible processes the principle of least action is no longer sufficient; for the principle of increase of entropy brings into the system of physics a wholly new element, foreign to the action principle, and which demands special mathematical treatment. The unidirectional course of a process in the attainment of a fixed final state is related to it.

I hope the foregoing considerations have sufficed to make clear to you that the distinction between reversible and irreversible processes is much broader than that between mechanical and electrical processes and that, therefore, this difference, with better right than any other, may be taken advantage of in classifying all physical processes, and that it may eventually play in the theoretical physics of the future the principal role.

However, the classification mentioned is in need of quite an essential improvement, for it cannot be denied that in the form set forth, the system of physics is still suffering from a strong dose of anthropomorphism. In the definition of irreversibility, as well as in that of entropy, reference is made to the possibility of carrying out in nature certain changes, and this means, fundamentally, nothing more than that the division of physical processes is made dependent upon the manipulative skill of man in the art of experimentation, which certainly does not always remain at a fixed stage, but is continually being more and more perfected. If, therefore, the distinction between reversible and irreversible processes is actually to have a lasting significance for all times, it must be essentially broadened and made independent of any reference to the capacities of mankind. How this may happen, I desire to state one week from tomorrow. The lecture of tomorrow will be devoted to the problem of bringing before you some of the most important of the great number of practical consequences following from the entropy principle.

2 THERMODYNAMIC STATES OF EQUILIBRIUM IN DILUTE SOLUTIONS

In the lecture of yesterday I sought to make clear the fact that the essential, and therefore the final division of all processes occurring in nature, is into reversible and irreversible processes, and the characteristic difference between these two kinds of processes, as I have further separated them, is that in irreversible processes the entropy increases, while in all reversible processes it remains constant. Today I am constrained to speak of some of the consequences of this law which will illustrate its rich fruitfulness. They have to do with the question of the laws of thermodynamic equilibrium. Since in nature the entropy can only increase, it follows that the state of a physical configuration which is completely isolated, and in which the entropy of the system possesses an absolute maximum, is necessarily a state of stable equilibrium, since for it no further change is possible. How deeply this law underlies all physical and chemical relations has been shown by no one better and more completely than by John Willard Gibbs, whose name, not only in America, but in the whole world will be counted among those of the most famous theoretical physicists of all times; to whom, to my sorrow, it is no longer possible for me to tender personally my respects. It would be gratuitous for me, here in the land of his activity, to expatiate fully on the progress of his ideas, but you will

perhaps permit me to speak in the lecture of today of some of
the important applications in which thermodynamic research,
based on Gibbs works, can be advanced beyond his results.

These applications refer to the theory of dilute solutions,
and we shall occupy ourselves today with these, while I show
you by a definite example what fruitfulness is inherent in thermodynamic theory. I shall first characterize the problem quite
generally. It has to do with the state of equilibrium of a material system of any number of arbitrary constituents in an
arbitrary number of phases, at a given temperature T and
given pressure p. If the system is completely isolated, and
therefore guarded against all external thermal and mechanical actions, then in any ensuing change the entropy of the
system will increase:

$$dS > 0.$$

But if, as we assume, the system stands in such relation to
its surroundings that in any change which the system under-
goes the temperature T and the pressure p are maintained
constant, as, for instance, through its introduction into a
calorimeter of great heat capacity and through loading with
a piston of fixed weight, the inequality would suffer a change
thereby. We must then take account of the fact that the
surrounding bodies also, e..g., the calorimetric liquid, will
be involved in the change. If we denote the entropy of the
surrounding bodies by S_0, then the following more general
equation holds:

$$dS + dS_0 > 0.$$

In this equation

$$dS_0 = -\frac{Q}{T},$$

if Q denote the heat which is given up in the change by the
surroundings to the system. On the other hand, if U denote
the energy, V the volume of the system, then, in accordance

with the first law of thermodynamics,

$$Q = dU + pdV.$$

Consequently, through substitution:

$$dS - \frac{dU + pdV}{T} > 0$$

or, since p and T are constant:

$$d\left(S - \frac{U + pV}{T}\right) > 0.$$

If, therefore, we put:

$$S - \frac{U + pV}{T} = \Phi, \tag{1}$$

then

$$d\Phi > 0,$$

and we have the general law, that in every isothermal-isobaric ($T = $ const., $p = $ const.) change of state of a physical system the quantity Φ increases. The absolutely stable state of equilibrium of the system is therefore characterized through the maximum of Φ:

$$\delta\Phi = 0. \tag{2}$$

If the system consist of numerous phases, then, because Φ, in accordance with (1), is linear and homogeneous in S, U and V, the quantity Φ referring to the whole system is the sum of the quantities Φ referring to the individual phases. If the expression for Φ is known as a function of the independent variables for each phase of the system, then, from equation (2), all questions concerning the conditions of stable equilibrium may be answered. Now, within limits, this is the case for dilute solutions. By "solution" in thermodynamics is meant each homogeneous phase, in whatever state of aggregation, which is composed of a series of different molecular

complexes, each of which is represented by a definite molecular number. If the molecular number of a given complex is
great with reference to all the remaining complexes, then the
solution is called dilute, and the molecular complex in question is called the solvent; the remaining complexes are called
the dissolved substances.

Let us now consider a dilute solution whose state is determined by the pressure p, the temperature T, and the molecular numbers n_0, n_1, n_2, n_3, $\cdots$, wherein the subscript zero
refers to the solvent. Then the numbers n_1, n_2, n_3, $\cdots$ are all
small with respect to n_0, and on this account the volume V
and the energy U are linear functions of the molecular numbers:

$$V = n_0 v_0 + n_1 v_1 + n_2 v_2 + \cdots,$$
$$U = n_0 u_0 + n_1 u_1 + n_2 u_2 + \cdots,$$

wherein the v's and u's depend upon p and T only.

From the general equation of entropy:

$$dS = \frac{dU + pdV}{T},$$

in which the differentials depend only upon changes in p and T,
and not in the molecular numbers, there results therefore:

$$dS = n_0 \frac{du_0 + pdv_0}{T} + n_1 \frac{du_1 + pdv_1}{T} + \cdots,$$

and from this it follows that the expressions multiplied by
n_0, n_1 $\cdots$, dependent upon p and T only, are complete differentials. We may therefore write:

$$\frac{du_0 + pdv_0}{T} = ds_0, \quad \frac{du_1 + pdv_1}{T} = ds_1, \quad \cdots \quad (3)$$

and by integration obtain:

$$S = n_0 s_0 + n_1 s_1 + n_2 s_2 + \cdots + C.$$

The constant C of integration does not depend upon p and T, but may depend upon the molecular numbers n_0, n_1, n_2, $\cdots$. In order to express this dependence generally, it suffices to know it for a special case, for fixed values of p and T. Now every solution passes, through appropriate increase of temperature and decrease of pressure, into the state of a mixture of ideal gases, and for this case the entropy is fully known, the integration constant being, in accordance with Gibbs:

$$C = -R(n_0 \log c_0 + n_1 \log c_1 + \cdots),$$

wherein R denotes the absolute gas constant and c_0, c_1, c_2, $\cdots$ denote the "molecular concentrations":

$$c_0 = \frac{n_0}{n_0 + n_1 + n_2 + \cdots}, \quad c_1 = \frac{n_1}{n_0 + n_1 + n_2 + \cdots}, \quad \cdots.$$

Consequently, quite in general, the entropy of a dilute solution is:

$$S = n_0(s_0 - R \log c_0) + n_1(s_1 - R \log c_1) + \cdots,$$

and, finally, from this it follows by substitution in equation (1) that:

$$\Phi = n_0(\varphi_0 - R \log c_0) + n_1(\varphi_1 - R \log c_1) + \cdots, \quad (4)$$

if we put for brevity:

$$\varphi_0 = s_0 - \frac{u_0 + pv_0}{T}, \quad \varphi_1 = s_1 - \frac{u_1 + pv_1}{T}, \quad \cdots \quad (5)$$

all of which quantities depend only upon p and T.

With the aid of the expression obtained for Φ we are enabled through equation (2) to answer the question with regard to thermodynamic equilibrium. We shall first find the general law of equilibrium and then apply it to a series of particularly interesting special cases.

Every material system consisting of an arbitrary number of homogeneous phases may be represented symbolically in the following way:

$$n_0 m_0, \; n_1 m_1, \; \cdots \mid n_0{}' m_0{}', \; n_1{}' m_1{}', \; \cdots \mid n_0{}'' m_0{}'', \; n_1{}'' m_1{}'', \; \cdots \mid \cdots .$$

Here the molecular numbers are denoted by n, the molecular weights by m, and the individual phases are separated from one another by vertical lines. We shall now suppose that each phase represents a dilute solution. This will be the case when each phase contains only a single molecular complex and therefore represents an absolutely pure substance; for then the concentrations of all the dissolved substances will be zero.

If now an isobaric-isothermal change in the system of such kind is possible that the molecular numbers

$$n_0, \; n_1, \; n_2, \; \cdots, \quad n_0{}', \; n_1{}', \; n_2{}', \; \cdots, \quad n_0{}'', \; n_1{}'', \; n_2{}'', \; \cdots$$

change simultaneously by the amounts

$$\delta n_0, \; \delta n_1, \; \delta n_2, \cdots, \quad \delta n_0{}', \; \delta n_1{}', \; \delta n_2{}', \cdots, \quad \delta n_0{}'', \; \delta n_1{}'', \; \delta n_2{}'', \cdots$$

then, in accordance with equation (2), equilibrium obtains with respect to the occurrence of this change if, when T and p are held constant, the function

$$\Phi + \Phi' + \Phi'' + \cdots$$

is a maximum, or, in accordance with equation (4):

$$\sum (\varphi_0 - R \log c_0)\delta n_0 + (\varphi_1 - R \log c_1)\delta n_1 + \cdots = 0$$

(the summation $\sum$ being extended over all phases of the system). Since we are only concerned in this equation with the ratios of the δn's, we put

$$\delta n_0 : \delta n_1 : \cdots : \delta n_0{}' : \delta n_1{}' : \cdots : \delta n_0{}'' : \delta n_1{}'' : \cdots$$
$$= \nu_0 : \nu_1 : \cdots : \nu_0{}' : \nu_1{}' : \cdots : \nu_0{}'' : \nu_1{}'' : \cdots,$$

wherein we are to understand by the simultaneously changing ν's, in the variation considered, simple integer positive or negative numbers, according as the molecular complex under consideration is formed or disappears in the change. Then the condition for equilibrium is:

$$\sum \nu_0 \log c_0 + \nu_1 \log c_1 + \cdots$$
$$= \frac{1}{R} \sum \nu_0 \varphi_0 + \nu_1 \varphi_1 + \cdots = \log K. \tag{6}$$

K and the quantities $\varphi_0, \varphi_1, \varphi_2, \cdots$ depend only upon p and T, and this dependence is to be found from the equations:

$$\frac{\partial \log K}{\partial p} = \frac{1}{R} \sum \nu_0 \frac{\partial \varphi_0}{\partial p} + \nu_1 \frac{\partial \varphi_1}{\partial p} + \cdots,$$
$$\frac{\partial \log K}{\partial T} = \frac{1}{R} \sum \nu_0 \frac{\partial \varphi_0}{\partial T} + \nu_1 \frac{\partial \varphi_1}{\partial T} + \cdots.$$

Now, in accordance with (5), for any infinitely small change of p and T:

$$d\varphi_0 = ds_0 - \frac{du_0 + pdv_0 + v_0 dp}{T} + \frac{u_0 + pv_0}{T^2} dT,$$

and consequently, from (3):

$$d\varphi_0 = \frac{u_0 + pv_0}{T^2} dT - \frac{v_0 dp}{T},$$

and hence:

$$\frac{\partial \varphi_0}{\partial p} = -\frac{v_0}{T}, \quad \frac{\partial \varphi_0}{\partial T} = \frac{u_0 + pv_0}{T^2}.$$

Similar equations hold for the other φ's, and therefore we get:

$$\frac{\partial \log K}{\partial p} = -\frac{1}{RT} \sum \nu_0 v_0 + \nu_1 v_1 + \cdots,$$
$$\frac{\partial \log K}{\partial T} = -\frac{1}{RT^2} \sum \nu_0 u_0 + \nu_2 u_2 + \cdots + p(\nu_0 v_0 + \nu_1 v_1 + \cdots)$$

or, more briefly:

$$\frac{\partial \log K}{\partial p} = -\frac{1}{RT}\Delta V, \quad \frac{\partial \log K}{\partial T} = \frac{\Delta Q}{RT^2}, \tag{7}$$

if ΔV denote the change in the total volume of the system and ΔQ the heat which is communicated to it from outside, during the isobaric isothermal change considered. We shall now investigate the import of these relations in a series of important applications.

2.1 Electrolytic Dissociation of Water

The system consists of a single phase:

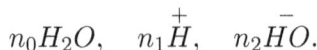

$$n_0 H_2 O, \quad n_1 \overset{+}{H}, \quad n_2 \overset{-}{H} O.$$

The transformation under consideration

$$\nu_0 : \nu_1 : \nu_2 = \delta n_0 : \delta n_1 : \delta n_2$$

consists in the dissociation of a molecule $H_2 O$ into a molecule $\overset{+}{H}$ and a molecule $\overset{-}{H} O$, therefore:

$$\nu_0 = -1, \quad \nu_1 = 1, \quad \nu_2 = 1.$$

Hence, in accordance with (6), for equilibrium:

$$- \log c_0 + \log c_1 + \log c_2 = \log K,$$

or, since $c_1 = c_2$ and $c_0 = 1$, approximately:

$$2 \log c_1 = \log K.$$

The dependence of the concentration c_1 upon the temperature now follows from (7):

$$2\frac{\partial \log c_1}{\partial T} = \frac{\Delta Q}{RT^2}.$$

ΔQ, the quantity of heat which it is necessary to supply for the dissociation of a molecule of H_2O into the ions $\overset{+}{H}$ and $\overset{-}{HO}$, is, in accordance with Arrhenius, equal to the heat of ionization in the neutralization of a strong univalent base and acid in a dilute aqueous solution, and, therefore, in accordance with the recent measurements of Wörmann,[1]

$$\Delta Q = 27{,}857 - 48.5T \text{ gr. cal.}$$

Using the number 1.985 for the ratio of the absolute gas constant R to the mechanical equivalent of heat, it follows that:

$$\frac{\partial \log c_1}{\partial T} = \frac{1}{2 \times 1.985}\left(\frac{27{,}857}{T^2} - \frac{48.5}{T}\right),$$

and by integration:

$$\overset{10}{\log} c_1 = -\frac{3047.3}{T} - 12.125 \overset{10}{\log} T + \text{const.}$$

This dependence of the degree of dissociation upon the temperature agrees very well with the measurements of the electric conductivity of water at different temperatures by Kohlrausch and Heydweiller, Noyes, and Lundén.

2.2 Dissociation of a Dissolved Electrolyte

Assume the system consists of an aqueous solution of acetic acid:

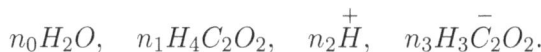

$$n_0 H_2O, \quad n_1 H_4C_2O_2, \quad n_2 \overset{+}{H}, \quad n_3 H_3\overset{-}{C_2}O_2.$$

The change under consideration consists in the dissociation of a molecule $H_4C_2O_2$ into its two ions, therefore

$$\nu_0 = 0, \quad \nu_1 = -1, \quad \nu_2 = 1, \quad \nu_3 = 1.$$

[1] Ad Heydweiller, Ann. d. Phys., 28, 506, 1909.

Hence, for the state of equilibrium, in accordance with (6):

$$-\log c_1 + \log c_2 + \log c_3 = \log K,$$

or, since $c_2 = c_3$:

$$\frac{c_2{}^2}{c_1} = K.$$

Now the sum $c_1 + c_2 = c$ is to be regarded as known, since the total number of the undissociated and dissociated acid molecules is independent of the degree of dissociation. Therefore c_1 and c_2 may be calculated from K and c. An experimental test of the equation of equilibrium is possible on account of the connection between the degree of dissociation and electrical conductivity of the solution. In accordance with the electrolytic dissociation theory of Arrhenius, the ratio of the molecular conductivity λ of the solution in any dilution to the molecular conductivity λ_∞ of the solution in infinite dilution is:

$$\frac{\lambda}{\lambda_\infty} = \frac{c_2}{c_1 + c_2} = \frac{c_2}{c},$$

since electric conduction is accounted for by the dissociated molecules only. It follows then, with the aid of the last equation, that:

$$\frac{\lambda^2 c}{\lambda_\infty - \lambda} = K \lambda_\infty = \text{const.}$$

With unlimited decreasing c, λ increases to λ_∞. This "law of dilution" for binary electrolytes, first enunciated by Ostwald, has been confirmed in numerous cases by experiment, as in the case of acetic acid.

Also, the dependence of the degree of dissociation upon the temperature is indicated here in quite an analogous manner to that in the example considered above, of the dissociation of water.

2.3 Vaporization or Solidification of a Pure Liquid

In equilibrium the system consists of two phases, one liquid, and one gaseous or solid:

$$n_0 m_0 \mid n_0' m_0'.$$

Each phase contains only a single molecular complex (the solvent), but the molecules in both phases do not need to be the same. Now, if a liquid molecule evaporates or solidifies, then in our notation

$$\nu_0 = -1, \quad \nu_0' = \frac{m_0}{m_0'}, \quad c_0 = 1, \quad c_0' = 1,$$

and consequently the condition for equilibrium, in accordance with (6), is:

$$0 = \log K. \tag{8}$$

Since K depends only upon p and T, this equation therefore expresses a definite relation between p and T: the law of dependence of the pressure of vaporization (or melting pressure) upon the temperature, or vice versa. The import of this law is obtained through the consideration of the dependence of the quantity K upon p and T. If we form the complete differential of the last equation, there results:

$$0 = \frac{\partial \log K}{\partial p} dp + \frac{\partial \log K}{\partial T} dT,$$

or, in accordance with (7):

$$0 = -\frac{\Delta V}{T} dp + \frac{\Delta Q}{T^2} dT.$$

If v_0 and v_0' denote the molecular volumes of the two phases, then:

$$\Delta V = \frac{m_0 v_0'}{m_0'} - v_0,$$

consequently:

$$\Delta Q = T \left(\frac{m_0 v_0'}{m_0'} - v_0 \right) \frac{dp}{dT},$$

or, referred to unit mass:

$$\frac{\Delta Q}{m_0} = T \left(\frac{v_0'}{m_0'} - \frac{v_0}{m_0} \right) \frac{dp}{dT},$$

the well-known formula of Carnot and Clapeyron.

2.4 The Vaporization or Solidification of a Solution of Non-Volatile Substances

Most aqueous salt solutions afford examples. The symbol of the system in this case is, since the second phase (gaseous or solid) contains only a single molecular complex:

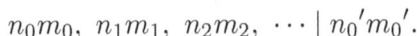

$$n_0 m_0, \; n_1 m_1, \; n_2 m_2, \; \cdots \mid n_0' m_0'.$$

The change is represented by:

$$\nu_0 = -1, \quad \nu_1 = 0, \quad \nu_2 = 0, \quad \cdots \quad \nu_0' = \frac{m_0}{m_0'},$$

and hence the condition of equilibrium, in accordance with (6), is:

$$- \log c_0 = \log K,$$

or, since to small quantities of higher order:

$$c_0 = \frac{n_0}{n_0 + n_1 + n_2 + \cdots} = 1 - \frac{n_1 + n_2 + \cdots}{n_0},$$

$$\frac{n_1 + n_2 + \cdots}{n_0} = \log K. \tag{9}$$

A comparison with formula (8), found in example III, shows that through the solution of a foreign substance there is involved in the total concentration a small proportionate departure from the law of vaporization or solidification which holds for the pure solvent. One can express this, either by saying: at a fixed pressure p, the boiling point or the freezing point T of the solution is different than that (T_0) for the pure solvent, or: at a fixed temperature[2] T the vapor pressure or solidification pressure p of the solution is different from that (p_0) of the pure solvent. Let us calculate the departure in both cases.

1. If T_0 be the boiling (or freezing temperature) of the pure solvent at the pressure p, then, in accordance with (8):

$$(\log K)_{T=T_0} = 0,$$

and by subtraction of (9) there results:

$$\log K - (\log K)_{T=T_0} = \frac{n_1 + n_2 + \cdots}{n_0}.$$

Now, since T is little different from T_0, we may write in place of this equation, with the aid of (7):

$$\frac{\partial \log K}{\partial T}(T - T_0) = \frac{\Delta Q}{R T_0^2}(T - T_0) = \frac{n_1 + n_2 + \cdots}{n_0},$$

and from this it follows that:

$$T - T_0 = \frac{n_1 + n_2 + \cdots}{n_0} \frac{R T_0^2}{\Delta Q}. \tag{10}$$

This is the law for the raising of the boiling point or for the lowering of the freezing point, first derived by van't Hoff: in the case of freezing ΔQ (the heat taken from the surroundings during the freezing of a liquid molecule) is negative. Since n_0 and ΔQ occur only as a product, it is not possible to

---

[2]EDITOR'S NOTE: This appears to be an error – in the original publication (p. 31) it is written "pressure T", but obviously it should be "temperature T."

infer anything from this formula with regard to the molecular number of the liquid solvent.

2. If p_0 be the vapor pressure of the pure solvent at the temperature T, then, in accordance with (8):

$$(\log K)_{p=p_0} = 0,$$

and by subtraction of (9) there results:

$$\log K - (\log K)_{p=p_0} = \frac{n_1 + n_2 + \cdots}{n_0}.$$

Now, since p and p_0 are nearly equal, with the aid of (7) we may write:

$$\frac{\partial \log K}{\partial p}(p - p_0) = -\frac{\Delta V}{RT}(p - p_0) = \frac{n_1 + n_2 + \cdots}{n_0},$$

and from this it follows, if ΔV be placed equal to the volume of the gaseous molecule produced in the vaporization of a liquid molecule:

$$\Delta V = \frac{m_0}{m_0'}\frac{RT}{p},$$

$$\frac{p_0 - p}{p} = \frac{m_0'}{m_0}\frac{n_1 + n_2 + \cdots}{n_0}.$$

This is the law of relative depression of the vapor pressure, first derived by van't Hoff. Since n_0 and m_0 occur only as a product, it is not possible to infer from this formula anything with regard to the molecular weight of the liquid solvent. Frequently the factor m_0'/m_0 is left out in this formula; but this is not allowable when m_0 and m_0' are unequal (as, e.g., in the case of water).

2.5 Vaporization of a Solution of Volatile Substances

(E.g.., a Sufficiently Dilute Solution of Propyl Alcohol in Water.)

The system, consisting of two phases, is represented by the following symbol:

$$n_0 m_0, \; n_1 m_1, \; n_2 m_2, \; \cdots \mid n_0' m_0', \; n_1' m_1', \; n_2' m_2', \; \cdots \,,$$

wherein, as above, the figure 0 refers to the solvent and the figures 1, 2, 3 $\cdots$ refer to the various molecular complexes of the dissolved substances. By the addition of primes in the case of the molecular weights $(m_0', m_1', m_2' \cdots)$ the possibility is left open that the various molecular complexes in the vapor may possess a different molecular weight than in the liquid.

Since the system here considered may experience various sorts of changes, there are also various conditions of equilibrium to fulfill, each of which relates to a definite sort of transformation. Let us consider first that change which consists in the vaporization of the solvent. In accordance with our scheme of notation, the following conditions hold:

$$\nu_0 = -1, \; \nu_1 = 0, \; \nu_2 = 0, \; \cdots \; \nu_0' = \frac{m_0}{m_0'}, \; \nu_1' = 0, \; \nu_2' = 0, \; \cdots \,,$$

and, therefore, the condition of equilibrium (6) becomes:

$$- \log c_0 + \frac{m_0}{m_0'} \log c_0' = \log K,$$

or, if one substitutes:

$$c_0 = 1 - \frac{n_1 + n_2 + \cdots}{n_0} \quad \text{and} \quad c_0' = 1 - \frac{n_1' + n_2' + \cdots}{n_0'},$$

$$\frac{n_1 + n_2 + \cdots}{n_0} - \frac{m_0}{m_0'} \frac{n_1' + n_2' + \cdots}{n_0'} = \log K.$$

If we treat this equation upon equation (9) as a model, there results an equation similar to (10):

$$T - T_0 = \left(\frac{n_1 + n_2 + \cdots}{n_0 m_0} - \frac{n_1' + n_2' + \cdots}{n_0' m_0'} \right) \frac{R T_0^2 m_0}{\Delta Q}.$$

Here ΔQ is the heat effect in the vaporization of one molecule of the solvent and, therefore, $\Delta Q / m_0$ is the heat effect in the vaporization of a unit mass of the solvent.

We remark, once more, that the solvent always occurs in the formula through the mass only, and not through the molecular number or the molecular weight, while, on the other hand, in the case of the dissolved substances, the molecular state is characteristic on account of their influence upon vaporization. Finally, the formula contains a generalization of the law of van't Hoff, stated above, for the raising of the boiling point, in that here in place of the number of dissolved molecules in the liquid, the difference between the number of dissolved molecules in unit mass of the liquid and in unit mass of the vapor appears. According as the unit mass of liquid or the unit mass of vapor contains more dissolved molecules, there results for the solution a raising or lowering of the boiling point; in the limiting case, when both quantities are equal, and the mixture therefore boils without changing, the change in boiling point becomes equal to zero. Of course, there are corresponding laws holding for the change in the vapor pressure.

Let us consider now a change which consists in the vaporization of a dissolved molecule. For this case we have in our notation

$$\nu_0 = 0, \ \nu_1 = -1, \ \nu_2 = 0 \ \cdots, \ \nu_0' = 0, \ \nu_1' = \frac{m_1}{m_1'}, \ \nu_2' = 0, \ \cdots$$

and, in accordance with (6), for the condition of equilibrium:

$$- \log c_1 + \frac{m_1}{m_1'} \log c_1' = \log K$$

or:

$$\frac{c_1'^{\frac{m_1}{m_1'}}}{c_1} = K.$$

This equation expresses the Nernst law of distribution. If the dissolved substance possesses in both phases the same molecular weight ($m_1 = m_1'$), then, in a state of equilibrium a fixed ratio of the concentrations c_1 and c_1' in the liquid and in the vapor exists, which depends only upon the pressure and temperature. But, if the dissolved substance polymerises somewhat in the liquid, then the relation demanded in the last equation appears in place of the simple ratio.

2.6 The Dissolved Substance only Passes over into the Second Phase

This case is in a certain sense a special case of the one preceding. To it belongs that of the solubility of a slightly soluble salt, first investigated by van't Hoff, e.g., succinic acid in water. The symbol of this system is:

$$n_0 H_2O, \; n_1 H_6 C_4 O_4 \mid n_0' H_6 C_4 O_4,$$

in which we disregard the small dissociation of the acid solution. The concentrations of the individual molecular complexes are:

$$c_0 = \frac{n_0}{n_0 + n_1}, \quad c_1 = \frac{n_1}{n_0 + n_1}, \quad c_0' = \frac{n_0'}{n_0'} = 1.$$

For the precipitation of solid succinic acid we have:

$$\nu_0 = 0, \quad \nu_1 = -1, \quad \nu_0' = 1,$$

and, therefore, from the condition of equilibrium (6):

$$- \log c_1 = \log K,$$

hence, from (7):

$$\Delta Q = -RT^2 \frac{\partial \log c_1}{\partial T}.$$

By means of this equation van't Hoff calculated the heat of solution ΔQ from the solubility of succinic acid at $0°$ and at

8.5° C. The corresponding numbers were 2.88 and 4.22 in an arbitrary unit. Approximately, then:

$$\frac{\partial \log c_1}{\partial T} = \frac{\overset{e}{\log} 4.22 - \overset{e}{\log} 2.88}{8.5} = 0.04494,$$

from which for $T = 273$:

$$\Delta Q = -1.98 \times 273^2 \times 0.04494 = -6,600 \text{ cal.,}$$

that is, in the precipitation of a molecule of succinic acid, $6,600$ cal. are given out to the surroundings. Berthelot found, however, through direct measurement, $6,700$ calories for the heat of solution.

The absorption of a gas also comes under this head, e.g. carbonic acid, in a liquid of relatively unnoticeable smaller vapor pressure, e.g., water at not too high a temperature. The symbol of the system is then

$$n_0 H_2 O, \; n_1 CO_2 \mid n_0' CO_2.$$

The vaporization of a molecule CO_2 corresponds to the values

$$\nu_0 = 0, \quad \nu_1 = -1, \quad \nu_0' = 1.$$

The condition of equilibrium is therefore again:

$$-\log c_1 = \log K,$$

i.e., at a fixed temperature and a fixed pressure the concentration c_1 of the gas in the solution is constant. The change of the concentration with p and T is obtained through substitution in equation (7). It follows from this that:

$$\frac{\partial \log c_1}{\partial p} = \frac{\Delta V}{RT}, \quad \frac{\partial \log c_1}{\partial T} = -\frac{\Delta Q}{RT^2}.$$

ΔV is the change in volume of the system which occurs in the isobaric-isothermal vaporization of a molecule of CO_2,

ΔQ the quantity of heat absorbed in the process from outside. Now, since ΔV represents approximately the volume of a molecule of gaseous carbonic acid, we may put approximately:

$$\Delta V = \frac{RT}{p},$$

and the equation gives:

$$\frac{\partial \log c_1}{\partial p} = \frac{1}{p},$$

which integrated, gives:

$$\log c_1 = \log p + \text{const.}, \quad c_1 = C\,p,$$

i.e., the concentration of the dissolved gas is proportional to the pressure of the free gas above the solution (law of Henry and Bunsen). The factor of proportionality C, which furnishes a measure of the solubility of the gas, depends upon the heat effect in quite the same manner as in the example previously considered.

A number of no less important relations are easily derived as by-products of those found above, e.g., the Nernst laws concerning the influence of solubility, the Arrhenius theory of isohydric solutions, etc. All such may be obtained through the application of the general condition of equilibrium (6). In conclusion, there is one other case that I desire to treat here. In the historical development of the theory this has played a particularly important rôle.

2.7   Osmotic Pressure

We consider now a dilute solution separated by a membrane (permeable with regard to the solvent but impermeable as regards the dissolved substance) from the pure solvent (in the same state of aggregation), and inquire as to the condition of equilibrium. The symbol of the system considered we may again take as

$$n_0 m_0, \quad n_1 m_1, \quad n_2 m_2, \quad \cdots \mid n_0' m_0.$$

The condition of equilibrium is also here again expressed by equation (6), valid for a change of state in which the temperature and the pressure in each phase is maintained constant. The only difference with respect to the cases treated earlier is this, that here, in the presence of a separating membrane between two phases, the pressure p in the first phase may be different from the pressure p' in the second phase, whereby by "pressure," as always, is to be understood the ordinary hydrostatic or manometric pressure.

The proof of the applicability of equation (6) is found in the same way as this equation was derived above, proceeding from the principle of increase of entropy. One has but to remember that, in the somewhat more general case here considered, the external work in a given change is represented by the sum $pdV + p'dV'$, where V and V' denote the volumes of the two individual phases, while before V denoted the total volume of all phases. Accordingly, we use, instead of (7), to express the dependence of the constant K in (6) upon the pressure:

$$\frac{\partial \log K}{\partial p} = -\frac{\Delta V}{RT}, \quad \frac{\partial \log K}{\partial p'} = -\frac{\Delta V'}{RT}. \tag{11}$$

We have here to do with the following change:

$$\nu_0 = -1, \quad \nu_1 = 0, \quad \nu_2 = 0, \quad \cdots, \quad \nu_0' = 1,$$

whereby is expressed, that a molecule of the solvent passes out of the solution through the membrane into the pure solvent. Hence, in accordance with (6):

$$-\log c_0 = \log K,$$

or, since

$$c_0 = 1 - \frac{n_1 + n_2 + \cdots}{n_0}, \quad \frac{n_1 + n_2 + \cdots}{n_0} = \log K.$$

Here K depends only upon T, p and p'. If a pure solvent were present upon both sides of the membrane, we should have $c_0 = 1$, and $p = p'$; consequently:

$$(\log K)_{p=p'} = 0,$$

and by subtraction of the last two equations:

$$\frac{n_1 + n_2 + \cdots}{n_0} = \log K - (\log K)_{p=p'} = \frac{\partial \log K}{\partial p}(p - p')$$

and in accordance with (11):

$$\frac{n_1 + n_2 + \cdots}{n_0} = -(p - p')\frac{\Delta V}{RT}.$$

Here ΔV denotes the change in volume of the solution due to the loss of a molecule of the solvent ($\nu_0 = -1$). Approximately then:

$$-\Delta V\, n_0 = V,$$

the volume of the whole solution, and

$$\frac{n_1 + n_2 + \cdots}{n_0} = (p - p')\frac{V}{RT}.$$

If we call the difference $p - p'$, the osmotic pressure of the solution, this equation contains the well known law of osmotic pressure, due to van't Hoff.

The equations here derived, which easily permit of multiplication and generalization, have, of course, for the most part not been derived in the ways described above, but have been derived, either directly from experiment, or theoretically from the consideration of special reversible isothermal cycles to which the thermodynamic law was applied, that in such a cyclic process not only the algebraic sum of the work produced and the heat produced, but that also each of these two quantities separately, is equal to zero (first lecture, p. 50). The employment of a cyclic process has the advantage over the procedure here proposed, that in it the connection between the directly measurable quantities and the requirements of the laws of thermodynamics succinctly appears in each case; but for each individual case a satisfactory cyclic process must be imagined, and one has not always the certain assurance that the thermodynamic realization of the cyclic process also actually supplies all the conditions of equilibrium. Furthermore, in the process of calculation certain terms of considerable weight frequently appear as empty ballast, since they disappear at the end in the summation over the individual phases of the process.

On the other hand, the significance of the process here employed consists therein, that the necessary and sufficient conditions of equilibrium for each individually considered case appear collectively in the single equation (6), and that they are derived collectively from it in a direct manner through an unambiguous procedure. The more complicated the systems considered are, the more apparent becomes the advantage of this method, and there is no doubt in my mind that in chemical circles it will be more and more employed, especially, since in general it is now the custom to deal directly with the energies, and not with cyclic processes, in the calculation of heat effects in chemical changes.

3 THE ATOMIC THEORY OF MATTER

The problem with which we shall be occupied in the present lecture is that of a closer investigation of the atomic theory of matter. It is, however, not my intention to introduce this theory with nothing further, and to set it up as something apart and disconnected with other physical theories, but I intend above all to bring out the peculiar significance of the atomic theory as related to the present general system of theoretical physics; for in this way only will it be possible to regard the whole system as one containing within itself the essential compact unity, and thereby to realize the principal object of these lectures.

Consequently it is self evident that we must rely on that sort of treatment which we have recognized in last week's lecture as fundamental. That is, the division of all physical processes into reversible and irreversible processes. Furthermore, we shall be convinced that the accomplishment of this division is only possible through the atomic theory of matter, or, in other words, that irreversibility leads of necessity to atomistics.

I have already referred at the close of the first lecture to the fact that in pure thermodynamics, which knows nothing of an atomic structure and which regards all substances as absolutely continuous, the difference between reversible and irreversible processes can only be defined in one way, which a priori carries a provisional character and does not withstand penetrating analysis. This appears immediately evident when one reflects that the purely thermodynamic definition of irreversibility which proceeds from the impossibility of the realization of certain changes in nature, as, e. g., the transformation of heat into work without compensation, has at the outset assumed a definite limit to man's mental capacity, while, however, such a limit is not indicated in reality. On the contrary: mankind is making every endeavor to press beyond the present boundaries of its capacity, and we hope that later on many things will be attained which, perhaps, many regard at present as impossible of accomplishment. Can it not happen then that a process, which up to the present has been regarded as irreversible, may be proved, through a new discovery or invention, to be reversible? In this case the whole structure of the second law would undeniably collapse, for the irreversibility of a single process conditions that of all the others.

It is evident then that the only means to assure to the second law real meaning consists in this, that the idea of irreversibility be made independent of any relationship to man and especially of all technical relations.

Now the idea of irreversibility harks back to the idea of entropy; for a process is irreversible when it is connected with an increase of entropy. The problem is hereby referred back to a proper improvement of the definition of entropy. In accordance with the original definition of Clausius, the entropy is measured by means of a certain reversible process, and the weakness of this definition rests upon the fact that many such reversible processes, strictly speaking all, are not capable of being carried out in practice. With some reason it may be objected that we have here to do, not with an actual process and an actual physicist, but only with ideal processes, so-called thought experiments, and with an ideal physicist who operates with all the experimental methods with absolute accuracy. But at this point the difficulty is encountered: How far do the physicist's ideal measurements of this sort suffice? It may be understood, by passing to the limit, that a gas is compressed by a pressure which is equal to the pressure of the gas, and is heated by a heat reservoir which possesses the same temperature as the gas, but, for example, that a saturated vapor shall be transformed through isothermal compression in a reversible manner to a liquid without at any time a part of the vapor being condensed, as in certain thermodynamic considerations is supposed, must certainly appear doubtful. Still more striking, however, is the liberty as regards thought experiments, which in physical chemistry is granted the theorist. With his semi-permeable membranes, which in reality are only realizable under certain special conditions and then only with a certain approximation, he separates in a reversible manner, not only all possible varieties of molecules, whether or not they are in stable or unstable conditions, but he also separates the oppositely charged ions from one another and from the undissociated molecules, and he is disturbed, neither by the enormous electrostatic forces which resist such a separation, nor by the circumstance that in reality, from the beginning of the separation, the molecules become in part dissociated while the ions in part again combine. But such ideal processes are necessary throughout in order to make possible the comparison of the entropy of the undissociated molecules with the entropy of the dissociated molecules; for the law of thermodynamic equilibrium does not permit in general of derivation in any other way, in case one wishes to retain pure thermodynamics as a basis. It must be considered remarkable that all these ingenious thought processes have so well found confirmation of their results in experience, as is shown by the examples considered by us in the last lecture.

If now, on the other hand, one reflects that in all these results every reference to the possibility of actually carrying out each ideal process has disappeared—there are certainly left relations between directly measurable quantities only, such as temperature, heat effect, concentration, etc.—the presumption forces itself upon one that perhaps the introduction as above of such ideal processes is at bottom a round-about method, and that the peculiar import of the principle of increase of entropy with all its consequences can be evolved from the original idea of irreversibility or, just as well, from the impossibility of perpetual motion of the second kind, just as the principle of conservation of energy has been evolved from the law of impossibility of perpetual motion of the first kind.

This step: to have completed the emancipation of the entropy idea from the experimental art of man and the elevation of the second law thereby to a real principle, was the scientific life's work of Ludwig Boltzmann. Briefly stated, it consisted in general of referring back the idea of entropy to the idea of probability. Thereby is also explained, at the same time, the significance of the above (p. 48) auxiliary term used by me; "preference" of nature for a definite state. Nature prefers the more probable states to the less probable, because in nature processes take place in the direction of greater probability. Heat goes from a body at higher temperature to a body at lower temperature because the state of equal temperature distribution is more probable than a state of unequal temperature distribution.

Through this conception the second law of thermodynamics is removed at one stroke from its isolated position, the mystery concerning the preference of nature vanishes, and the entropy principle reduces to a well understood law of the calculus of probability.

The enormous fruitfulness of so "objective" a definition of entropy for all domains of physics I shall seek to demonstrate in the following lectures. But today we have principally to do with the proof of its admissibility; for on closer consideration we shall immediately perceive that the new conception of entropy at once introduces a great number of questions, new requirements and difficult problems. The first requirement is the introduction of the atomic hypothesis into the system of physics. For, if one wishes to speak of the probability of a physical state, i. e., if he wishes to introduce the probability for a given state as a definite quantity into the calculation, this can only be brought about, as in cases of all probability calculations, by referring the state back to a variety of possibilities; i. e., by considering a finite number of a priori equally likely configurations (complexions) through each of which the state considered may be realized. The greater the number of complexions, the greater is the probability of the state. Thus, e. g., the probability of throwing a total of four with two ordinary six-sided dice is found through counting the complexions by which the throw with a total of four may be realized. Of these there are three complexions:

> with the first die, 1, with the second die, 3,
> with the first die, 2, with the second die, 2,
> with the first die, 3, with the second die, 1.

On the other hand, the throw of two is only realized through a single complexion. Therefore, the probability of throwing a total of four is three times as great as the probability of throwing a total of two.

Now, in connection with the physical state under consideration, in order to be able to differentiate completely from one another the complexions realizing it, and to associate it with a definite reckonable number, there is obviously no other means than to regard it as made up of numerous discrete homogeneous elements—for in perfectly continuous systems there exist no reckonable elements—and hereby the atomistic view is made a fundamental requirement. We have, therefore, to regard all bodies in nature, in so far as they possess an entropy, as constituted of atoms, and we therefore arrive in physics at the same conception of matter as that which obtained in chemistry for so long previously.

But we can immediately go a step further yet. The conclusions reached hold, not only for thermodynamics of material bodies, but also possess complete validity for the processes of heat radiation, which are thus referred back to the second law of thermodynamics. That radiant heat also possesses an entropy follows from the fact that a body which emits radiation into a surrounding diathermanous medium experiences a loss of heat and, therefore, a decrease of entropy. Since the total entropy of a physical system can only increase, it follows that one part of the entropy of the whole system, consisting of the body and the diathermanous medium, must be contained in the radiated heat. If the entropy of the radiant heat is to be referred back to the notion of probability, we are forced, in a similar way as above, to the conclusion that for radiant heat the atomic conception possesses a definite meaning. But, since radiant heat is not directly connected with matter, it follows that this atomistic conception relates, not to matter, but only to energy, and hence, that in heat radiation certain energy elements play an essential rôle. Even though this conclusion appears so singular and even though in many circles today vigorous objection is strongly urged against it, in the long run physical research will not be able to withhold its sanction from it, and the less, since it is confirmed by experience in quite a satisfactory manner. We shall return to this point in the lectures on heat radiation. I desire here only to mention that the novelty involved by the introduction of atomistic conceptions into the theory of heat radiation is by no means so revolutionary as, perhaps, might appear at the first glance. For there is, in my opinion at least, nothing which makes necessary the consideration of the heat processes in a complete vacuum as atomic, and it suffices to seek the atomistic features at the source of radiation, i. e., in those processes which have their play in the centres of emission and absorption of radiation. Then the Maxwellian electrodynamic differential equations can retain completely their validity for the vacuum, and, besides, the discrete elements of heat radiation are relegated exclusively to a domain which is still very mysterious and where there is still present plenty of room for all sorts of hypotheses.

Returning to more general considerations, the most important question comes up as to whether, with the introduction of atomistic conceptions and with the reference of entropy to probability, the content of the principle of increase of entropy is exhaustively comprehended, or whether still further physical hypotheses are required in order to secure the full import of that principle. If this important question had been settled at the time of the introduction of the atomic theory into thermodynamics, then the atomistic views would surely have been spared a large number of conceivable misunderstandings and justifiable attacks. For it turns out, in fact—and our further considerations will confirm this conclusion—that there has as yet nothing been done with atomistics which in itself requires much more than an essential generalization, in order to guarantee the validity of the second law.

We must first reflect that, in accordance with the central idea laid down in the first lecture (p. 37), the second law must possess validity as an objective physical law, independently of the individuality of the physicist. There is nothing to hinder us from imagining a physicist—we shall designate him a "microscopic" observer—whose senses are so sharpened that he is able to recognize each individual atom and to follow it in its motion. For this observer each atom moves exactly in accordance with the elementary laws which general dynamics lays down for it, and these laws allow, so far as we know, of an inverse performance of every process. Accordingly, here again the question is neither one of probability nor of entropy and its increase. Let us imagine, on the other hand, another observer, designated a "macroscopic" observer, who regards an ensemble of atoms as a homogeneous gas, say, and consequently applies the laws of thermodynamics to the mechanical and thermal processes within it. Then, for such an observer, in accordance with the second law, the process in general is an irreversible process. Would not now the first observer be justified in saying: "The reference of the entropy to probability has its origin in the fact that irreversible processes ought to be explained through reversible processes. At any rate, this procedure appears to me in the highest degree dubious. In any case, I declare each change of state which takes place in the ensemble of atoms designated a gas, as reversible, in opposition to the macroscopic observer." There is not the slightest thing, so far as I know, that one can urge against the validity of these statements. But do we not thereby place ourselves in the painful position of the judge who declared in a trial the correctness of the position of each separately of two contending parties and then, when a third contends that only one of the parties could emerge from the process victorious, was obliged to declare him also correct? Fortunately we find ourselves in a more favorable position. We can certainly mediate between the two parties without its being necessary for one or the other to give up his principal point of view. For closer consideration shows that the whole controversy rests upon a misunderstanding—a new proof of how necessary it is before one begins a controversy to come to an understanding with his opponent concerning the subject of the quarrel. Certainly, a given change of state cannot be both reversible and irreversible. But the one observer connects a wholly different idea with the phrase "change of state" than the other. What is then, in general, a change of state? The state of a physical system cannot well be otherwise defined than as the aggregate of all those physical quantities, through whose instantaneous values the time changes of the quantities, with given boundary conditions, are uniquely determined. If we inquire now, in accordance with the import of this definition,

of the two observers as to what they understand by the state
of the collection of atoms or the gas considered, they will give
quite different answers. The microscopic observer will mention those quantities which determine the position and the
velocities of all the individual atoms. There are present in
the simplest case, namely, that in which the atoms may be
considered as material points, six times as many quantities
as atoms, namely, for each atom the three coordinates and
the three velocity components, and in the case of combined
molecules, still more quantities. For him the state and the
progress of a process is then first determined when all these
various quantities are individually given. We shall designate
the state defined in this way the "micro-state." The macroscopic observer, on the other hand, requires fewer data. He
will say that the state of the homogeneous gas considered by
him is determined by the density, the visible velocity and the
temperature at each point of the gas, and he will expect that,
when these quantities are given, their time variations and,
therefore, the progress of the process, to be completely determined in accordance with the two laws of thermo-dynamics,
and therefore accompanied by an increase in entropy. In this
connection he can call upon all the experience at his disposal,
which will fully confirm his expectation. If we call this state
the "macro-state," it is clear that the two laws: "the micro-
changes of state are reversible" and "the macro-changes of
state are irreversible," lie in wholly different domains and, at
any rate, are not contradictory.

But now how can we succeed in bringing the two observers
to an understanding? This is a question whose answer is obviously of fundamental significance for the atomic theory. First
of all, it is easy to see that the macro-observer reckons only
with mean values; for what he calls density, visible velocity
and temperature of the gas are, for the micro-observer, certain mean values, statistical data, which are derived from the
space distribution and from the velocities of the atoms in an

appropriate manner. But the micro-observer cannot operate with these mean values alone, for, if these are given at one instant of time, the progress of the process is not determined throughout; on the contrary: he can easily find with given mean values an enormously large number of individual values for the positions and the velocities of the atoms, all of which correspond with the same mean values and which, in spite of this, lead to quite different processes with regard to the mean values. It follows from this of necessity that the micro-observer must either give up the attempt to understand the unique progress, in accordance with experience, of the macroscopic changes of state—and this would be the end of the atomic theory—or that he, through the introduction of a special physical hypothesis, restrict in a suitable manner the manifold of micro-states considered by him. There is certainly nothing to prevent him from assuming that not all conceivable micro-states are realizable in nature, and that certain of them are in fact thinkable, but never actually realized. In the formularization of such a hypothesis, there is of course no point of departure to be found from the principles of dynamics alone; for pure dynamics leaves this case undetermined. But on just this account any dynamical hypothesis, which involves nothing further than a closer specification of the micro-states realized in nature, is certainly permissible. Which hypothesis is to be given the preference can only be decided through comparison of the results to which the different possible hypotheses lead in the course of experience.

In order to limit the investigation in this way, we must obviously fix our attention only upon all imaginable configurations and velocities of the individual atoms which are compatible with determinate values of the density, the velocity and the temperature of the gas, or in other words: we must consider all the micro-states which belong to a determinate macro-state, and must investigate the various kinds of processes which follow in accordance with the fixed laws of dynamics from the different micro-states. Now, precise calculation has in every case always led to the important result that an enormously large number of these different micro-processes relate to one and the same macro-process, and that only proportionately few of the same, which are distinguished by quite special exceptional conditions concerning the positions and the velocities of neighboring atoms, furnish exceptions. Furthermore, it has also shown that one of the resulting macro-processes is that which the macroscopic observer recognizes, so that it is compatible with the second law of thermodynamics.

Here, manifestly, the bridge of understanding is supplied. The micro-observer needs only to assimilate in his theory the physical hypothesis that all those special cases in which special exceptional conditions exist among the neighboring configurations of interacting atoms do not occur in nature, or, in other words, that the micro-states are in elementary disorder. Then the uniqueness of the macroscopic process is assured and with it, also, the fulfillment of the principle of increase of entropy in all directions.

Therefore, it is not the atomic distribution, but rather the hypothesis of elementary disorder, which forms the real kernel of the principle of increase of entropy and, therefore, the preliminary condition for the existence of entropy. Without elementary disorder there is neither entropy nor irreversible process.[1] Therefore, a single atom can never possess an en-

---

[1] To those physicists who, in spite of all this, regard the hypothesis of elementary disorder as gratuitous or as incorrect, I wish to refer the simple fact that in every calculation of a coefficient of friction, of diffusion, or of heat conduction, from molecular considerations, the notion of elementary disorder is employed, whether tacitly or otherwise, and that it is therefore essentially more correct to stipulate this condition instead of ignoring or concealing it. But he who regards the hypothesis of elementary disorder as self-evident, should be reminded that, in accordance with a law of H. Poincaré, the precise investigation concerning the foundation of which would here lead us too far, the assumption of

tropy; for we cannot speak of disorder in connection with it. But with a fairly large number of atoms, say 100 or 1,000, the matter is quite different. Here, one can certainly speak of a disorder, in case that the values of the coordinates and the velocity components are distributed among the atoms in accordance with the laws of accident. Then it is possible to calculate the probability for a given state. But how is it with regard to the increase of entropy? May we assert that the motion of 100 atoms is irreversible? Certainly not; but this is only because the state of 100 atoms cannot be defined in a thermodynamic sense, since the process does not proceed in a unique manner from the standpoint of a macro-observer, and this requirement forms, as we have seen above, the foundation and preliminary condition for the definition of a thermodynamic state.

If one therefore asks: How many atoms are at least necessary in order that a process may be considered irreversible?, the answer is: so many atoms that one may form from them definite mean values which define the state in a macroscopic sense. One must reflect that to secure the validity of the principle of increase of entropy there must be added to the condition of elementary disorder still another, namely, that the number of the elements under consideration be sufficiently large to render possible the formation of definite mean values. The second law has a meaning for these mean values only; but for them, it is quite exact, just as exact as the law of the calculus of probability, that the mean value, so far as it may be defined, of a sufficiently large number of throws with a six-sided die, is $3\frac{1}{2}$.

These considerations are, at the same time, capable of throwing light upon questions such as the following: Does the principle of increase of entropy possess a meaning for the

---

this hypothesis for all times is unwarranted for a closed space with absolutely smooth walls,—an important conclusion, against which can only be urged the fact that absolutely smooth walls do not exist in nature.

so-called Brownian molecular movement of a suspended particle? Does the kinetic energy of this motion represent useful work or not? The entropy principle is just as little valid for a single suspended particle as for an atom, and therefore is not valid for a few of them, but only when there is so large a number that definite mean values can be formed. That one is able to see the particles and not the atoms makes no material difference; because the progress of a process does not depend upon the power of an observing instrument. The question with regard to useful work plays no rôle in this connection; strictly speaking, this possesses, in general, no objective physical meaning. For it does not admit of an answer without reference to the scheme of the physicist or technician who proposes to make use of the work in question. The second law, therefore, has fundamentally nothing to do with the idea of useful work (cf. first lecture, p. 46).

But, if the entropy principle is to hold, a further assumption is necessary, concerning the various disordered elements,—an assumption which tacitly is commonly made and which we have not previously definitely expressed. It is, however, not less important than those referred to above. The elements must actually be of the same kind, or they must at least form a number of groups of like kind, e. g., constitute a mixture in which each kind of element occurs in large numbers. For only through the similarity of the elements does it come about that order and law can result in the larger from the smaller. If the molecules of a gas be all different from one another, the properties of a gas can never show so simple a law-abiding behavior as that which is indicated by thermodynamics. In fact, the calculation of the probability of a state presupposes that all complexions which correspond to the state are a priori equally likely. Without this condition one is just as little able to calculate the probability of a given state as, for instance, the probability of a given throw with dice whose sides are unequal in size. In summing up we may therefore say:

the second law of thermodynamics in its objective physical conception, freed from anthropomorphism, relates to certain mean values which are formed from a large number of disordered elements of the same kind.

The validity of the principle of increase of entropy and of the irreversible progress of thermodynamic processes in nature is completely assured in this formularization. After the introduction of the hypothesis of elementary disorder, the microscopic observer can no longer confidently assert that each process considered by him in a collection of atoms is reversible; for the motion occurring in the reverse order will not always obey the requirements of that hypothesis. In fact, the motions of single atoms are always reversible, and thus far one may say, as before, that the irreversible processes appear reduced to a reversible process, but the phenomenon as a whole is nevertheless irreversible, because upon reversal the disorder of the numerous individual elementary processes would be eliminated. Irreversibility is inherent, not in the individual elementary processes themselves, but solely in their irregular constitution. It is this only which guarantees the unique change of the macroscopic mean values.

Thus, for example, the reverse progress of a frictional process is impossible, in that it would presuppose elementary arrangement of interacting neighboring molecules. For the collisions between any two molecules must thereby possess a certain distinguishing character, in that the velocities of two colliding molecules depend in a definite way upon the place at which they meet. In this way only can it happen that in collisions like directed velocities ensue and, therefore, visible motion.

Previously we have only referred to the principle of elementary disorder in its application to the atomic theory of matter. But it may also be assumed as valid, as I wish to indicate at this point, on quite the same grounds as those holding in the case of matter, for the theory of radiant heat. Let us consider, e. g., two bodies at different temperatures between which exchange of heat occurs through radiation. We can in this case also imagine a microscopic observer, as opposed to the ordinary macroscopic observer, who possesses insight into all the particulars of electromagnetic processes which are connected with emission and absorption, and the propagation of heat rays. The microscopic observer would declare the whole process reversible because all electrodynamic processes can also take place in the reverse direction, and the contradiction may here be referred back to a difference in definition of the state of a heat ray. Thus, while the macroscopic observer completely defines a monochromatic ray through direction, state of polarization, color, and intensity, the microscopic observer, in order to possess a complete knowledge of an electromagnetic state, necessarily requires the specification of all the numerous irregular variations of amplitude and phase to which the most homogeneous heat ray is actually subject. That such irregular variations actually exist follows immediately from the well known fact that two rays of the same color never interfere, except when they originate in the same source of light. But until these fluctuations are given in all particulars, the micro-observer can say nothing with regard to the progress of the process. He is also unable to specify whether the exchange of heat radiation between the two bodies leads to a decrease or to an increase of their difference in temperature. The principle of elementary disorder first furnishes the adequate criterion of the tendency of the radiation process, i. e., the warming of the colder body at the expense of the warmer, just as the same principle conditions the irreversibility of exchange of heat through conduction. However, in the two cases compared, there is indicated an essential difference in the kind of the disorder. While in heat conduction the disordered elements may be represented as associated with the various molecules, in heat radiation there are the numerous vibration periods, connected with a heat ray, among which the energy of radiation is irregularly distributed. In other words: the disorder among the molecules is a material one, while in heat radiation it is one of energy distribution. This is the most important difference between the two kinds of disorder; a common feature exists as regards the great number of uncoordinated elements required. Just as the entropy of a body is defined as a function of the macroscopic state, only when the body contains so many atoms that from them definite mean values may be formed, so the entropy principle only possesses a meaning with regard to a heat ray when the ray comprehends so many periodic vibrations, i. e., persists for so long a time, that a definite mean value for the intensity of the ray may be obtained from the successive irregular fluctuating amplitudes.

Now, after the principle of elementary disorder has been introduced and accepted by us as valid throughout nature, the fundamental question arises as to the calculation of the probability of a given state, and the actual derivation of the entropy therefrom. From the entropy all the laws of thermodynamic states of equilibrium, for material substances, and also for energy radiation, may be uniquely derived. With regard to the connection between entropy and probability, this is inferred very simply from the law that the probability of two independent configurations is represented by the product of the individual probabilities:

$$W = W_1 W_2,$$

while the entropy S is represented by the sum of the individual entropies:

$$S = S_1 + S_2.$$

Accordingly, the entropy is proportional to the logarithm of the probability:

$$S = k \log W. \tag{12}$$

k is a universal constant. In particular, it is the same for atomic as for radiation configurations, for there is nothing to prevent us assuming that the configuration designated by 1 is atomic, while that designated by 2 is a radiation configuration. If k has been calculated, say with the aid of radiation measurements, then k must have the same value for atomic processes. Later we shall follow this procedure, in order to utilize the laws of heat radiation in the kinetic theory of gases. Now, there remains, as the last and most difficult part of the problem, the calculation of the probability W of a given physical configuration in a given macroscopic state. We shall treat today, by way of preparation for the quite general problem to follow, the simple problem: to specify the probability of a given state for a single moving material point, subject to given conservative forces. Since the state depends upon 6 variables: the 3 generalized coordinates φ_1, φ_2, φ_3, and the three corresponding velocity components $\dot{\varphi}_1$, $\dot{\varphi}_2$, $\dot{\varphi}_3$, and since all possible values of these 6 variables constitute a continuous manifold, the probability sought is, that these 6 quantities shall lie respectively within certain infinitely small intervals, or, if one thinks of these 6 quantities as the rectilinear orthogonal coordinates of a point in an ideal six-dimensional space, that this ideal "state point" shall fall within a given, infinitely small "state domain." Since the domain is infinitely small, the probability will be proportional to the magnitude of the domain and therefore proportional to

$$\int d\varphi_1 \cdot d\varphi_2 \cdot d\varphi_3 \cdot d\dot{\varphi}_1 \cdot d\dot{\varphi}_2 \cdot d\dot{\varphi}_3.$$

But this expression cannot serve as an absolute measure of the probability, because in general it changes in magnitude with the time, if each state point moves in accordance with the laws of motion of material points, while the probability of a state which follows of necessity from another must be the same for the one as the other. Now, as is well known, another integral quite similarly formed, may be specified in place of the one above, which possesses the special property of not changing in value with the time. It is only necessary to employ, in addition to the general coordinates φ_1, φ_2, φ_3, the three so-called momenta ψ_1, ψ_2, ψ_3, in place of the three velocities $\dot{\varphi}_1$, $\dot{\varphi}_2$, $\dot{\varphi}_3$ as the determining coordinates of the state. These are defined in the following way:

$$\psi_1 = \left(\frac{\partial H}{\partial \dot{\varphi}_1}\right)_\phi, \quad \psi_2 = \left(\frac{\partial H}{\partial \dot{\varphi}_2}\right)_\phi, \quad \psi_3 = \left(\frac{\partial H}{\partial \dot{\varphi}_3}\right)_\phi,$$

wherein H denotes the kinetic potential (Helmholz). Then, in Hamiltonian form, the equations of motion are:

$$\dot{\psi}_1 = \frac{d\psi_1}{dt} = -\left(\frac{\partial E}{\partial \varphi_1}\right)_\psi, \quad \cdots, \quad \dot{\varphi}_1 = \frac{d\varphi_1}{dt} = \left(\frac{\partial E}{\partial \psi_1}\right)_\phi, \quad \cdots,$$

(E is the energy), and from these equations follows the "condition of incompressibility":

$$\frac{\partial \dot{\varphi}_1}{\partial \varphi_1} + \frac{\partial \dot{\psi}_1}{\partial \psi_1} + \cdots = 0.$$

Referring to the six-dimensional space represented by the coordinates φ_1, φ_2, φ_3, ψ_1, ψ_2, ψ_3, this equation states that the magnitude of an arbitrarily chosen state domain, viz.:

$$\int d\varphi_1 \, d\varphi_2 \, d\varphi_3 \, d\psi_1 \, d\psi_2 \, d\psi_3$$

does not change with the time, when each point of the domain changes its position in accordance with the laws of motion of material points. Accordingly, it is made possible to take the magnitude of this domain as a direct measure for the probability that the state point falls within the domain.

From the last expression, which can be easily generalized for the case of an arbitrary number of variables, we shall calculate later the probability of a thermodynamic state, for the case of radiant energy as well as that for material substances.

4 The Equation of State for a Monatomic Gas

My problem today is to utilize the general fundamental laws concerning the concept of irreversibility, which we established in the lecture of yesterday, in the solution of a definite problem: the calculation of the entropy of an ideal monatomic gas in a given state, and the derivation of all its thermodynamic properties. The way in which we have to proceed is prescribed for us by the general definition of entropy:

$$S = k \log W. \tag{13}$$

The chief part of our problem is the calculation of W for a given state of the gas, and in this connection there is first required a more precise investigation of that which is to be understood as the state of the gas. Obviously, the state is to be taken here solely in the sense of the conception which we have called macroscopic in the last lecture. Otherwise, a state would possess neither probability nor entropy. Furthermore, we are not allowed to assume a condition of equilibrium for the gas. For this is characterized through the further special condition that the entropy for it is a maximum. Thus, an unequal distribution of density may exist in the gas; also, there may be present an arbitrary number of different currents, and in general no kind of equality between the various velocities of the molecules is to be assumed. The velocities, as the coordinates of the molecules, are rather to be taken a priori as quite arbitrarily given, but in order that the state, considered in a macroscopic sense, may be assumed as known, certain mean values of the densities and the velocities must exist. Through these mean values the state from a macroscopic standpoint is completely characterized.

The conditions mentioned will all be fulfilled if we consider the state as given in such manner that the number of molecules in a sufficiently small macroscopic space, but which, however, contains a very large number of molecules, is given, and furthermore, that the (likewise great) number of these molecules is given, which are found in a certain macroscopically small velocity domain, i. e., whose velocities lie within certain small intervals. If we call the coordinates x, y, z, and the velocity components $\dot{x}$, $\dot{y}$, $\dot{z}$, then this number will be proportional to[1]

$$dx \cdot dy \cdot dz \cdot d\dot{x} \cdot d\dot{y} \cdot d\dot{z} = \sigma.$$

It will depend, besides, upon a finite factor of proportionality which may be an arbitrarily given function $f(x, y, z, \dot{x}, \dot{y}, \dot{z})$ of the coordinates and the velocities, and which has only the one condition to fulfill that

$$\sum f \cdot \sigma = N, \tag{14}$$

where N denotes the total number of molecules in the gas. We are now concerned with the calculation of the probability W of that state of the gas which corresponds to the arbitrarily given distribution function f.

---

[1] We can call σ a "macro-differential" in contradistinction to the micro-differentials which are infinitely small with reference to the dimensions of a molecule. I prefer this terminology for the discrimination between "physical" and "mathematical" differentials in spite of the inelegance of phrasing, because the macro-differential is also just as much mathematical as physical and the micro-differential just as much physical as mathematical.

The probability that a given molecule possesses such coordinates and such velocities that it lies within the domain σ is expressed, in accordance with the final result of the previous lecture, by the magnitude of the corresponding elementary domain:

$$d\varphi_1 \cdot d\varphi_2 \cdot d\varphi_3 \cdot d\psi_1 \cdot d\psi_2 \cdot d\psi_3,$$

therefore, since here

$$\varphi_1 = x, \quad \varphi_2 = y, \quad \varphi_3 = z, \quad \psi_1 = m\dot{x}, \quad \psi_2 = m\dot{y}, \quad \psi_3 = m\dot{z},$$

(m the mass of a molecule) by

$$m^3\sigma.$$

Now we divide the whole of the six dimensional "state domain" containing all the molecules into suitable equal elementary domains of the magnitude $m^3\sigma$. Then the probability that a given molecule fall in a given elementary domain is equally great for all such domains. Let P denote the number of these equal elementary domains. Next, let us imagine as many dice as there are molecules present, i. e., N, and each die to be provided with P equal sides. Upon these P sides we imagine numbers $1, 2, 3, \cdots$ to P, so that each of the P sides indicates a given elementary domain. Then each throw with the N dice corresponds to a given state of the gas, while the number of dice which show a given number corresponds to the molecules which lie in the elementary domain considered. In accordance with this, each single die can indicate with the same probability each of the numbers from 1 to P, corresponding to the circumstance that each molecule may fall with equal probability in any one of the P elementary domains. The probability W sought, of the given state of the molecules, corresponds, therefore, to the number of different kinds of throws (complexions) through which is realized the given distribution f. Let us take, e. g., N equal to 10 molecules (dice) and $P = 6$ elementary domains (sides) and let us imagine the state so given that there are

3 molecules in 1st elementary domain
4 molecules in 2d elementary domain
0 molecules in 3d elementary domain
1 molecule in 4th elementary domain
0 molecules in 5th elementary domain
2 molecules in 6th elementary domain,

then this state, e. g., may be realized through a throw for which the 10 dice indicate the following numbers:

1st	2d	3d	4th	5th	6th	7th	8th	9th	10th
2	6	2	1	1	2	6	2	1	4.

$$(15)$$

Under each of the characters representing the ten dice stands the number which the die indicates in the throw. In fact,

3 dice show the figure 1
4 dice show the figure 2
0 dice show the figure 3
1 die shows the figure 4
0 dice show the figure 5
2 dice show the figure 6.

The state in question may likewise be realized through many other complexions of this kind. The number sought of all possible complexions is now found through consideration of the number series indicated in (15). For, since the number of molecules (dice) is given, the number series contains a fixed number of elements $(10 = N)$. Furthermore, since the number of molecules falling in an elementary domain is given, each number, in all permissible complexions, appears equally often in the series. Finally, each change of the number configuration conditions a new complexion. The number of possible complexions or the probability W of the given state is therefore equal to the number of possible permutations with repetition under the conditions mentioned. In the simple example chosen, in accordance with a well known formula, the probability is

$$\frac{10!}{3!\,4!\,0!\,1!\,0!\,2!} = 12,600.$$

Therefore, in the general case:

$$W = \frac{N!}{\prod(f \cdot \sigma)!}.$$

The sign $\prod$ denotes the product extended over all of the P elementary domains.

From this there results, in accordance with equation (13), for the entropy of the gas in the given state:

$$S = k \log N! - k\sum \log(f \cdot \sigma)!.$$

The summation is to be extended over all domains σ. Since $f \cdot \sigma$ is a large quantity, Stirling's formula may be employed for its factorial, which for a large number n is expressed by:

$$n! = \left(\frac{n}{e}\right)^n \sqrt{2\pi n}, \tag{16}$$

therefore, neglecting unimportant terms:

$$\log n! = n(\log n - 1);$$

and hence:

$$S = k \log N! - k\sum f\sigma(\log[f \cdot \sigma] - 1),$$

or, if we note that σ and $N = \sum f\sigma$ remain constant in all changes of state:

$$S = \text{const} - k\sum f \cdot \log f \cdot \sigma. \tag{17}$$

This quantity is, to the universal factor $(-k)$, the same as that which L. Boltzmann denoted by H, and which he showed to vary in one direction only for all changes of state.

In particular, we will now determine the entropy of a gas in a state of equilibrium, and inquire first as to that form of the law of distribution which corresponds to thermodynamic equilibrium. In accordance with the second law of thermodynamics, a state of equilibrium is characterized by the condition that with given values of the total volume V and the total energy E, the entropy S assumes its maximum value. If we assume the total volume of the gas

$$V = \int dx \cdot dy \cdot dz,$$

and the total energy

$$E = \frac{m}{2} \sum (\dot{x}^2 + \dot{y}^2 + \dot{z}^2) f \sigma \qquad (18)$$

as given, then the condition:

$$\delta S = 0$$

must hold for the state of equilibrium, or, in accordance with (17):

$$\sum (\log f + 1) \cdot \delta f \cdot \sigma = 0, \qquad (19)$$

wherein the variation δf refers to an arbitrary change in the law of distribution, compatible with the given values of N, V and E.

Now we have, on account of the constancy of the total number of molecules N, in accordance with (14):

$$\sum \delta f \cdot \sigma = 0$$

and, on account of the constancy of the total energy, in accordance with (18):

$$\sum (\dot{x}^2 + \dot{y}^2 + \dot{z}^2) \cdot \delta f \cdot \sigma = 0.$$

Consequently, for the fulfillment of condition (19) for all permissible values of δf, it is sufficient and necessary that

$$\log f + \beta(\dot{x}^2 + \dot{y}^2 + \dot{z}^2) = \text{const},$$

or:

$$f = \alpha e^{-\beta(\dot{x}^2 + \dot{y}^2 + \dot{z}^2)},$$

wherein α and β are constants. In the state of equilibrium, therefore, the space distribution of molecules is uniform, i. e., independent of x, y, z, and the distribution of velocities is the well known Maxwellian distribution.

The values of the constants α and β are to be found from those of N, V and E. For the substitution of the value found for f in (14) leads to:

$$N = V\alpha \left(\frac{\pi}{\beta}\right)^{\frac{3}{2}},$$

and the substitution of f in (18) leads to:

$$E = \tfrac{3}{4} V m \frac{\alpha}{\beta} \left(\frac{\pi}{\beta}\right)^{\frac{3}{2}}.$$

From these equations it follows that:

$$\alpha = \frac{N}{V} \cdot \left(\frac{3mN}{4\pi E}\right)^{\frac{3}{2}}, \quad \beta = \frac{3mN}{4E},$$

and hence finally, in accordance with (17), the expression for the entropy S of the gas in a state of equilibrium with given values for N, V and E is:

$$S = \text{const} + kN(\tfrac{3}{2} \log E + \log V). \tag{20}$$

The additive constant contains terms in N and m, but not in E and V.

The determination of the entropy here carried out permits now the specification directly of the complete thermodynamic behavior of the gas, viz., of the equation of state, and of the values of the specific heats. From the general thermodynamic definition of entropy:

$$dS = \frac{dE + pdV}{T}$$

are obtained the partial differential quotients of S with regard to E and V respectively:

$$\left(\frac{\partial S}{\partial E}\right)_V = \frac{1}{T}, \quad \left(\frac{\partial S}{\partial V}\right)_E = \frac{p}{T}.$$

Consequently, with the aid of (20):

$$\left(\frac{\partial S}{\partial E}\right)_V = \frac{3}{2}\frac{kN}{E} = \frac{1}{T}, \tag{21}$$

and

$$\left(\frac{\partial S}{\partial V}\right)_E = \frac{kN}{V} = \frac{p}{T}. \tag{22}$$

The second of these equations:

$$p = \frac{kNT}{V}$$

contains the laws of Boyle, Gay Lussac and Avogadro, the latter because the pressure depends only upon the number N, and not upon the constitution of the molecules. Writing it in the ordinary form:

$$p = \frac{RnT}{V},$$

where n denotes the number of gram molecules or mols of the gas, referred to $O_2 = 32g$, and R the absolute gas constant:

$$R = 8.315 \times 10^7 \frac{\text{erg}}{\text{deg}},$$

we obtain by comparison:

$$k = \frac{Rn}{N}. \tag{23}$$

If we denote the ratio of the mol number to the molecular number by ω, or, what is the same thing, the ratio of the molecular mass to the mol mass:

$$\omega = \frac{n}{N},$$

and hence:

$$k = \omega R. \tag{24}$$

From this, if ω is given, we can calculate the universal constant k, and conversely.

The equation (21) gives:

$$E = \tfrac{3}{2}kNT. \tag{25}$$

Now since the energy of an ideal gas is given by:

$$E = Anc_v T,$$

wherein c_v denotes in calories the heat capacity at constant volume of a mol, A the mechanical equivalent of heat:

$$A = 4.19 \times 10^7 \, \frac{\text{erg}}{\text{cal}},$$

it follows that:

$$c_v = \frac{3kN}{2An},$$

and, having regard to (23), we obtain:

$$c_v = \frac{3}{2}\frac{R}{A} = 3.0, \tag{26}$$

the mol heat in calories of any monatomic gas at constant volume.

For the mol heat c_p at constant pressure we have from the first law of thermodynamics

$$c_p - c_v = \frac{R}{A},$$

and, therefore, having regard to (26):

$$c_p = 5, \quad \frac{c_p}{c_v} = \tfrac{5}{3},$$

a known result for monatomic gases.

The mean kinetic energy L of a molecule is obtained from (25):

$$L = \frac{E}{N} = \tfrac{3}{2}kT. \tag{27}$$

You notice that we have derived all these relations through the identification of the mechanical with the thermodynamic expression for the entropy, and from this you recognize the fruitfulness of the method here proposed.

But a method can first demonstrate fully its usefulness when we utilize it, not only to derive laws which are already known, but when we apply it in domains for whose investigation there at present exist no other methods. In this connection its application affords various possibilities. Take the case of a monatomic gas which is not sufficiently attenuated to have the properties of the ideal state; there are here, as pointed out by J. D. van der Waals, two things to consider: (1) the finite size of the atoms, (2) the forces which act among the atoms. Taking account of these involves a change in the value of the probability and in the energy of the gas as well, and, so far as can now be shown, the corresponding change in the conditions for thermodynamic equilibrium leads to an equation of state which agrees with that of van der Waals. Certainly there is here a rich field for further investigations, of greater promise when experimental tests of the equation of state exist in larger number.

Another important application of the theory has to do with heat radiation, with which we shall be occupied the coming week. We shall proceed then in a similar way as here, and shall be able from the expression for the entropy of radiation to derive the thermodynamic properties of radiant heat.

Today we will refer briefly to the treatment of polyatomic gases. I have previously, upon good grounds, limited the treatment to monatomic molecules; for up to the present real difficulties appear to stand in the way of a generalization, from the principles employed by us, to include polyatomic molecules; in fact, if we wish to be quite frank, we must say that a satisfactory mechanical theory of polyatomic gases has not yet been found. Consequently, at present we do not know to what place in the system of theoretical physics to assign the processes within a molecule—the intra-molecular processes. We are obviously confronted by puzzling problems. A noteworthy and much discussed beginning was, it is true, made by Boltzmann, who introduced the most plausible assumption that for intra-molecular processes simple laws of the same kind hold as for the motion of the molecules themselves, *i. e.*, the general equations of dynamics. It is easy then, in fact, to proceed to the proof that for a monatomic gas the molecular heat c_v must be greater than 3 and that consequently, since the difference $c_p - c_v$ is always equal to 2, the ratio is

$$\frac{c_p}{c_v} = \frac{c_v + 2}{c_v} < \tfrac{5}{3}.$$

This conclusion is completely confirmed by experience. But this in itself does not confirm the assumption of Boltzmann; for, indeed, the same conclusion is reached very simply from the assumption that there exists intra-molecular energy which increases with the temperature. For then the molecular heat of a polyatomic gas must be greater by a corresponding amount than that of a monatomic gas.

Nevertheless, up to this point the Boltzmann theory never leads to contradiction with experience. But so soon as one seeks to draw special conclusions concerning the magnitude of the specific heats hazardous difficulties arise; I will refer to only one of them. If one assumes the Hamiltonian equations of mechanics as applicable to intra-molecular motions, he arrives of necessity at the law of "uniform distribution of energy," which asserts that under certain conditions, not essential to consider here, in a thermodynamic state of equilibrium the total energy of the gas is distributed uniformly among all the individual energy phases corresponding to the independent variables of state, or, as one may briefly say; the same amount of energy is associated with every independent variable of state. Accordingly, the mean energy of motion of the molecules $\frac{1}{2}kT$, corresponding to a given direction in space, is the same as for any other direction, and, moreover, the same for all the different kinds of molecules, and ions; also for all suspended particles (dust) in the gas, of whatever size, and, furthermore, the same for all kinds of motions of the constituents of a molecule relative to its centroid. If one now reflects that a molecule commonly contains, so far as we know, quite a large number of different freely moving constituents, certainly, that a normal molecule of a monatomic gas, e. g., mercury, possesses numerous freely moving electrons, then, in accordance with the law of uniform energy distribution, the intra-molecular energy must constitute a much larger fraction of the whole specific heat of the gas, and therefore c_p/c_v must turn out much smaller, than is consistent with the measured values. Thus, e. g., for an atom of mercury, in accordance with the measured value of $c_p/c_v = 5/3$, no part whatever of the heat added may be assigned to the intra-molecular energy. Boltzmann and others, in order to eliminate this contradiction, have fixed upon the possibility that, within the time of observation of the specific heats, the vibrations of the constituents (of a molecule) do not change appreciably with respect to one another, and come later with their progressive motion so slowly into heat equilibrium that this process is no longer capable of detection through observation. Up to now no such delay in the establishment of a state of equilibrium has been observed. Perhaps it would be productive of results if in delicate measurements special attention were paid the question as to whether observations which take a longer time lead to a greater value of the mol-heat, or, what comes to the same thing, a smaller value of c_p/c_v, than observations lasting a shorter time.

If one has been made mistrustful through these considerations concerning the applicability of the law of uniform energy distribution to intra-molecular processes, the mistrust is accentuated upon the inclusion of the laws of heat radiation. I shall make mention of this in a later lecture.

When we pass from stable atoms to the unstable atoms of radioactive substances, the principles following from the kinetic gas theory lose their validity completely. For the striking failure of all attempts to find any influence of temperature upon radioactive phenomena shows us that an application here of the law of uniform energy distribution is certainly not warranted. It will, therefore, be safest meanwhile to offer no definite conjectures with regard to the nature and the laws of these noteworthy phenomena, and to leave this field for further development to experimental research alone, which, I may say, with every day throws new light upon the subject.

106

5 HEAT RADIATION. ELECTRODYNAMIC THEORY

Last week I endeavored to point out that we find in the atomic theory a complete explanation for the whole content of the two laws of thermodynamics, if we, with Boltzmann, define the entropy by the probability, and I have further shown, in the example of an ideal monatomic gas, how the calculation of the probability, without any additional special hypothesis, enables us not only to find the properties of gases known from thermodynamics, but also to reach conclusions which lie essentially beyond those of pure thermodynamics. Thus, e. g., the law of Avogadro in pure thermodynamics is only a definition, while in the kinetic theory it is a necessary consequence; furthermore, the value of c_v, the mol-heat of a gas, is completely undetermined by pure thermodynamics, but from the kinetic theory it is of equal magnitude for all monatomic gases and, in fact, equal to 3, corresponding to our experimental knowledge. Today and tomorrow we shall be occupied with the application of the theory to radiant heat, and it will appear that we reach in this apparently quite isolated domain conclusions which a thorough test shows are compatible with experience. Naturally, we take as a basis the electromagnetic theory of heat radiation, which regards the rays as electro-magnetic waves of the same kind as light rays.

We shall utilize the time today in developing in bold outline the important consequences which follow from the electro-magnetic theory for the characteristic quantities of heat radiation, and tomorrow seek to answer, through the calculation of the entropy, the question concerning the dependence of these quantities upon the temperature, as was done last week for ideal gases. Above all, we are concerned here with the determination of those quantities which at any place in a medium traversed by heat rays determine the state of the radiant heat. The state of radiation at a given place will not be represented by a vector which is determined by three components; for the energy flowing in a given direction is quite independent of that flowing in any other direction. In order to know the state of radiation, we must be able to specify, moreover, the energy which in the time dt flows through a surface element $d\sigma$ for every direction in space. This will be proportional to the magnitude of $d\sigma$, to the time dt, and to the cosine of the angle ϑ which the direction considered makes with the normal to $d\sigma$. But the quantity to be multiplied by $d\sigma \cdot dt \cdot \cos \vartheta$ will not be a finite quantity; for since the radiation through any point of $d\sigma$ passes in all directions, therefore the quantity will also depend upon the magnitude of the solid angle $d\Omega$, which we shall assume as the same for all points of $d\sigma$. In this manner we obtain for the energy which in the time dt flows through the surface element $d\sigma$ in the direction of the elementary cone $d\Omega$, the expression:

$$K d\sigma dt \cdot \cos \vartheta \cdot d\Omega. \tag{28}$$

K is a positive function of place, of time and of direction, and is for unpolarized light of the following form:

$$K = 2 \int_0^\infty \mathfrak{K}_\nu d\nu \tag{29}$$

where ν denotes the frequency of a color of wave length λ and whose velocity of propagation is q:

$$\nu = \frac{q}{\lambda},$$

and $\mathfrak{K}_\nu$ denotes the corresponding intensity of spectral radiation of the plane polarized light.

From the value of K is to be found the space density of radiation ϵ, i. e., the energy of radiation contained in unit volume. The point 0 in question forms the centre of a sphere whose radius r we take so small that in the distance r no appreciable absorption of radiation takes place. Then each element $d\sigma$ of the surface of the sphere furnishes, by virtue of the radiation traversing the same, the following contribution to the radiation density at 0:

$$\frac{d\sigma \cdot dt \cdot K \cdot d\Omega}{r^2 d\Omega \cdot q dt} = \frac{d\sigma \cdot K}{r^2 q}.$$

For the radiation cone of solid angle $d\Omega$ proceeding from a point of $d\sigma$ in the direction toward 0 has at the distance r from $d\sigma$ the cross-section $r^2 d\Omega$ and the energy passing in the time dt through this cross-section distributes itself along the distance $q dt$. By integration over all of the surface elements $d\sigma$ we obtain the total space density of radiation at 0:

$$\epsilon = \int \frac{d\sigma K}{r^2 q} = \frac{1}{q} \int K d\Omega,$$

wherein $d\Omega$ denotes the solid angle of an elementary cone whose vertex is 0. For uniform radiation we obtain:

$$\epsilon = \frac{4\pi K}{q} = \frac{8\pi}{q} \cdot \int_0^\infty \mathfrak{K}_\nu d\nu. \tag{30}$$

The production of radiant heat is a consequence of the act of emission, and its destruction is the result of absorption. Both processes, emission and absorption, have their origin only in material particles, atoms or electrons, not at the geometrical bounding surface; although one frequently says, for the sake of brevity, that a surface element emits or absorbs. In reality a surface element of a body is a place of entrance for the radiation falling upon the body from without and which is to be absorbed; or a place of exit for the radiation emitted from within the body and passing through the surface in the outward direction. The capacity for emission and the capacity for absorption of an element of a body depend only upon its own condition and not upon that of the surrounding elements. If, therefore, as we shall assume in what follows, the state of the body varies only with the temperature, then the capacity for emission and the capacity for absorption of the body will also vary only with the temperature. The dependence upon the temperature can of course be different for each wave length.

We shall now introduce that result following from the second law of thermodynamics which will serve us as a basis in all subsequent considerations: "a system of bodies at rest of arbitrary nature, form and position, which is surrounded by a fixed shell impervious to heat, passes in the course of time from an arbitrarily chosen initial state to a permanent state in which the temperature of all bodies of the system is the same." This is the thermodynamic state of equilibrium in which the entropy of the system, among all those values which it may assume compatible with the total energy specified by the initial conditions, has a maximum value. Let us now apply this law to a single homogeneous isotropic medium which is of great extent in all directions of space and which, as in all cases subsequently considered, is surrounded by a fixed shell, perfectly reflecting as regards heat rays. The medium possesses for each frequency ν of the heat rays a finite capacity for emission and a finite capacity for absorption. Let us consider, now, such regions of the medium as are very far removed from the surface. Here the influence of the surface will be in any case vanishingly small, because no rays from the surface reach these regions, and on account of the homogeneity and isotropy of the medium we must conclude that the heat radiation is in thermodynamic equilibrium everywhere and has the same properties in all directions, so that $\mathfrak{K}_\nu$, the specific intensity of radiation of a plane polarized ray, is independent of the frequency ν, of the azimuth of polarization, of the direction of the ray, and of location. Thus, there will correspond to each diverging bundle of rays in an elementary cone $d\Omega$, proceeding from a surface element $d\sigma$, an exactly equal bundle oppositely directed, within the same elemental cone converging toward the surface element. This law retains its validity, as a simple consideration shows, right up to the surface of the medium. For in thermodynamic equilibrium each ray must possess exactly the same intensity as that of the directly opposite ray, otherwise, more energy would flow in one direction than in the opposite direction. Let us fix our attention upon a ray proceeding inwards from the surface, this must have the same intensity as that of the directly opposite ray coming from within, and from this it follows immediately that the state of radiation of the medium at all points on the surface is the same as that within. The nature of the bounding surface and the spacial extent of the medium are immaterial, and in a stationary state of radiation $\mathfrak{K}_\nu$ is completely determined by the nature of the medium for each temperature.

This law suffers a modification, however, in the special case that the medium is absolutely diathermanous for a definite frequency ν. It is then clear that the capacity for absorption and also that for emission must be zero, because otherwise no stationary state of radiation could exist, i. e., a medium emits no color which it does not absorb. But equilibrium can then obviously exist for every intensity of radiation of the frequency considered, i. e., $\mathfrak{K}_\nu$ is now undetermined and cannot be found without knowledge of the initial conditions. An important example of this is furnished by an absolute vacuum, which is diathermanous for all frequencies. In a complete vacuum thermodynamic equilibrium can therefore exist for each arbitrary intensity of radiation and for each frequency, i. e., for each arbitrary distribution of the spectral energy. From a general thermodynamic point of view this indeterminateness of the properties of thermodynamic states of equilibrium is explained through the presence of numerous different relative maxima of the entropy, as in the case of a vapor which is in a state of supersaturation. But among all the different maxima there is a special maximum, the absolute, which indicates stable equilibrium. In fact, we shall see that in a diathermanous medium for each temperature there exists a quite definite intensity of radiation, which is designated as the stable intensity of radiation of the frequency ν considered. But for the present we shall assume for all frequencies a finite capacity for absorption and for emission.

We consider now two homogeneous isotropic media in thermodynamic equilibrium separated from each other by a plane surface. Since the equilibrium will not be disturbed if one imagines for the moment the surface of separation between the two substances to be replaced by a surface quite non-transparent to heat radiation, all of the foregoing laws hold for each of the two substances individually. Let the specific intensity of radiation of frequency ν, polarized in any arbitrary plane within the first substance (the upper in Fig. 1)[1], be $\mathfrak{K}_\nu$ and that within the second substance $\mathfrak{K}_\nu{}'$ (we shall in general designate with a dash those quantities which refer to the second substance). Both quantities $\mathfrak{K}_\nu$ and $\mathfrak{K}_\nu{}'$, besides depending upon the temperature and the frequency, depend only upon the nature of the two substances, and, in fact, these values of the intensity of radiation hold quite up to the boundary surface between the substances, and are therefore independent of the properties of this surface.

Each ray from the first medium is split into two rays at the boundary surface: the reflected and the transmitted. The directions of these two rays vary according to the angle of

---

[1]From my lectures upon the theory of heat radiation (Leipzig, J. A. Barth), wherein are to be found the details of the above somewhat abbreviated calculations.

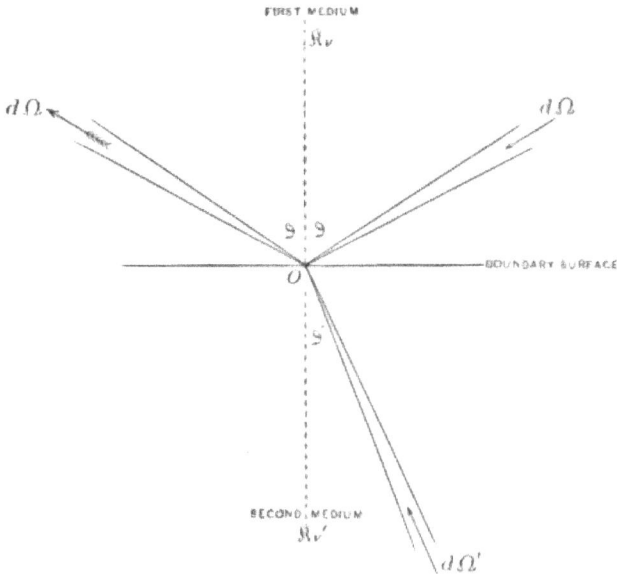

Fig. 1

incidence and the color of the incident ray, and, in addition, the intensity varies according to its polarization. If we denote by ρ (the reflection coefficient) the amount of the reflected energy of radiation and consequently by $1 - \rho$ the amount of transmitted energy with respect to the incident energy, then ρ depends upon the angle of incidence, upon the frequency and upon the polarization of the incident ray. Similar remarks hold for ρ', the reflection coefficient for a ray from the second medium, upon meeting the boundary surface.

Now the energy of a monochromatic plane polarized ray of frequency ν proceeding from an element $d\sigma$ of the boundary surface within the elementary cone $d\Omega$ in a direction toward the first medium (see the feathered arrow at the left in Fig. 1) is for the time dt, in accordance with (28) and (29):

$$dt \cdot d\sigma \cdot \cos\vartheta \cdot d\Omega \cdot \mathfrak{K}_\nu d\nu, \tag{31}$$

where
$$d\Omega = \sin\vartheta d\vartheta d\varphi. \tag{32}$$

This energy is furnished by the two rays which, approaching the surface from the first and the second medium respectively, are reflected and transmitted respectively at the surface element $d\sigma$ in the same direction. (See the unfeathered arrows. The surface element $d\sigma$ is indicated only by the point 0.) The first ray proceeds in accordance with the law of reflection within the symmetrically drawn elementary cone $d\Omega$: the second approaches the surface within the elementary cone

$$d\Omega' = \sin\vartheta' d\vartheta' d\varphi', \tag{33}$$

where, in accordance with the law of refraction,

$$\varphi' = \varphi \quad \text{and} \quad \frac{\sin\vartheta}{\sin\vartheta'} = \frac{q}{q'}. \tag{34}$$

We now assume that the ray is either polarized in the plane of incidence or perpendicular to this plane, and likewise for the two radiations out of whose energies it is composed. The radiation coming from the first medium and reflected from $d\sigma$ contributes the energy:

$$\rho \cdot dt \cdot d\sigma \cos\vartheta \cdot d\Omega \cdot \Re_\nu d\nu, \tag{35}$$

and the radiation coming from the second medium and transmitted through $d\sigma$ contributes the energy:

$$(1 - \rho') \cdot dt \cdot d\sigma \cos\vartheta' \cdot d\Omega' \cdot \Re_\nu' d\nu. \tag{36}$$

The quantities dt, $d\sigma$, ν, and $d\nu$ are here written without the accent, since they have the same values in both media.

Adding the expressions (35) and (36) and placing the sum equal to the expression (31), we obtain:

$$\rho \cos\vartheta d\Omega \Re_\nu + (1 - \rho') \cos\vartheta' d\Omega' \Re_\nu' = \cos\vartheta d\Omega \Re_\nu.$$

Now, in accordance with (34):

$$\frac{\cos \vartheta d\vartheta}{q} = \frac{\cos \vartheta' d\vartheta'}{q'},$$

and further, taking note of (32) and (33):

$$d\Omega' \cos \vartheta' = d\Omega \cos \vartheta \cdot \frac{q'^2}{q^2},$$

and it follows that:

$$\rho \mathfrak{K}_\nu + (1 - \rho') \frac{q'^2}{q^2} \mathfrak{K}_\nu' = \mathfrak{K}_\nu$$

or:

$$\frac{\mathfrak{K}_\nu}{\mathfrak{K}_\nu'} \cdot \frac{q^2}{q'^2} = \frac{1 - \rho'}{1 - \rho}.$$

In the last equation the quantity on the left is independent of the angle of incidence ϑ and of the kind of polarization, consequently the quantity upon the right side must also be independent of these quantities. If one knows the value of these quantities for a single angle of incidence and for a given kind of polarization, then this value is valid for all angles of incidence and for all polarizations. Now, in the particular case that the rays are polarized at right angles to the plane of incidence and meet the bounding surface at the angle of polarization,

$$\rho = 0 \quad \text{and} \quad \rho' = 0.$$

Then the expression on the right will be equal to 1, and therefore it is in general equal to 1, and we have always:

$$\rho = \rho', \quad q^2 \mathfrak{K}_\nu = q'^2 \mathfrak{K}_\nu'. \tag{37}$$

The first of these two relations, which asserts that the coefficient of reflection is the same for both sides of the boundary surface, constitutes the special expression of a general reciprocal law, first announced by Helmholz, whereby the loss of intensity which a ray of given color and polarization suffers on its path through any medium in consequence of reflection, refraction, absorption, and dispersion is exactly equal to the loss of intensity which a ray of corresponding intensity, color and polarization suffers in passing over the directly opposite path. It follows immediately from this that the radiation meeting a boundary surface between two media is transmitted or reflected equally well from both sides, for every color, direction and polarization.

The second relation, (37), brings into connection the radiation intensities originating in both substances. It asserts that in thermodynamic equilibrium the specific intensities of radiation of a definite frequency in both media vary inversely as the square of the velocities of propagation, or directly as the squares of the refractive indices. We may therefore write

$$q^2 \mathfrak{K}_\nu = F(\nu, T),$$

wherein F denotes a universal function depending only upon ν and T, the discovery of which is one of the chief problems of the theory.

Let us fix our attention again on the case of a diathermanous medium. We saw above that in a medium surrounded by a non-transparent shell which for a given color is diathermanous equilibrium can exist for any given intensity of radiation of this color. But it follows from the second law that, among all the intensities of radiation, a definite one, namely, that corresponding to the absolute maximum of the total entropy of the system, must exist, which characterizes the absolutely stable equilibrium of radiation. We now see that this indeterminateness is eliminated by the last equation, which asserts that in thermodynamic equilibrium the product $q^2 \mathfrak{K}_\nu$ is a universal function. For it results immediately therefrom that there is a definite value of $\mathfrak{K}_\nu$ for every diathermanous medium which is thus differentiated from all other values. The physical meaning of this value is derived directly from a consideration of the way in which this equation was derived: it is that intensity of radiation which exists in the diathermanous medium when it is in thermodynamic equilibrium while in contact with a given absorbing and emitting medium. The volume and the form of the second medium is immaterial; in particular, the volume may be taken arbitrarily small.

For a vacuum, the most diathermanous of all media, in which the velocity of propagation $q = c$ is the same for all rays, we can therefore express the following law: The quantity

$$\mathfrak{K}_\nu = \frac{1}{c^2} F(\nu, T) \tag{38}$$

denotes that intensity of radiation which exists in any complete vacuum when it is in a stationary state as regards exchange of radiation with any absorbing and emitting substance, whose amount may be arbitrarily small. This quantity $\mathfrak{K}_\nu$ regarded as a function of ν gives the so-called normal energy spectrum.

Let us consider, therefore, a vacuum surrounded by given emitting and absorbing bodies of uniform temperature. Then, in the course of time, there is established therein a normal energy radiation $\mathfrak{K}_\nu$ corresponding to this temperature. If now ρ_ν be the reflection coefficient of a wall for the frequency ν, then of the radiation $\mathfrak{K}_\nu$ falling upon the wall, the part $\rho_\nu \mathfrak{K}_\nu$ will be reflected. On the other hand, if we designate by E_ν the emission coefficient of the wall for the same frequency ν, the total radiation proceeding from the wall will be:

$$\rho_\nu \mathfrak{K}_\nu + E_\nu = \mathfrak{K}_\nu,$$

since each bundle of rays possesses in a stationary state the intensity $\mathfrak{K}_\nu$. From this it follows that:

$$\mathfrak{K}_\nu = \frac{E_\nu}{1 - \rho_\nu}, \tag{39}$$

i. e., the ratio of the emission coefficient E_ν to the capacity for absorption $(1 - \rho_\nu)$ of a given substance is the same for all substances and equal to the normal intensity of radiation for each frequency (Kirchoff). For the special case that ρ_ν is equal to 0, i. e., that the wall shall be perfectly black, we have:

$$\mathfrak{K}_\nu = E_\nu,$$

that is, the normal intensity of radiation is exactly equal to the emission coefficient of a black body. Therefore the normal radiation is also called "black radiation." Again, for any given body, in accordance with (39), we have:

$$E_\nu < \mathfrak{K}_\nu,$$

i. e., the emission coefficient of a body in general is smaller than that of a black body. Black radiation, thanks to W. Wien and O. Lummer, has been made possible of measurement, through a small hole bored in the wall bounding the space considered.

We proceed now to the treatment of the problem of determining the specific intensity $\mathfrak{K}_\nu$ of black radiation in a vacuum, as regards its dependence upon the frequency ν and the temperature T. In the treatment of this problem it will be necessary to go further than we have previously done into those processes which condition the production and destruction of heat rays; that is, into the question regarding the act of emission and that of absorption. On account of the complicated nature of these processes and the difficulty of bringing some of the details into connection with experience, it is certainly quite out of the question to obtain in this manner any reliable results if the following law cannot be utilized as a dependable guide in this domain: a vacuum surrounded by reflecting walls in which arbitrary emitting and absorbing bodies are distributed in any given arrangement assumes in the course of time the stationary state of black radiation, which is completely determined by a single parameter, the temperature, and which, in particular, does not depend upon the number, the properties and the arrangement of the bodies. In the investigation of the properties of the state of black radiation the nature of the bodies which are supposed to be in the vacuum is therefore quite immaterial, and it is certainly immaterial whether such bodies actually exist anywhere in nature, so long as their existence and their properties are compatible throughout with the laws of electrodynamics and of thermodynamics. As soon as it is possible to associate with any given special kind and arrangement of emitting and absorbing bodies a state of radiation in the surrounding vacuum which is characterized by absolute stability, then this state can be no other than that of black radiation. Making use of the freedom furnished by this law, we choose among all the emitting and absorbing systems conceivable, the most simple, namely, a single oscillator at rest, consisting of two poles charged with equal quantities of electricity of opposite sign which are movable relative to each other in a fixed straight line, the axis of the oscillator. The state of the oscillator is completely determined by its moment, $f(t)$; i. e., by the product of the electric charge of the pole on the positive side of the axis into the distance between the poles, and by its differential quotient with regard to the time:

$$\frac{df(t)}{dt} = \dot{f}(t).$$

The energy of the oscillator is of the following simple form:

$$U = \tfrac{1}{2}Kf^2 + \tfrac{1}{2}L\dot{f}^2, \tag{40}$$

wherein K and L denote positive constants which depend upon the nature of the oscillator in some manner into which we need not go further at this time.

If, in the vibrations of the oscillator, the energy U remain absolutely constant, we should have: $dU = 0$ or:

$$Kf(t) + L\ddot{f}(t) = 0,$$

and from this there results, as a general solution of the differential equation, a pure periodic vibration:

$$f = C\cos(2\pi\nu_0 t - \vartheta),$$

wherein C and ϑ denote the integration constants and ν_0 the number of vibrations per unit of time:

$$\nu_0 = \frac{1}{2\pi}\sqrt{\frac{K}{L}}. \tag{41}$$

Such an oscillator vibrating periodically with constant energy would neither be influenced by the electromagnetic field surrounding it, nor would it exert any external actions due to radiation. It could therefore have no sort of influence on the heat radiation in the surrounding vacuum.

In accordance with the theory of Maxwell, the energy of vibration U of the oscillator by no means remains constant in general, but an oscillator by virtue of its vibrations sends out spherical waves in all directions into the surrounding field and, in accordance with the principle of conservation of energy, if no actions from without are exerted upon the oscillator, there must necessarily be a loss in the energy of vibration and, therefore, a damping of the amplitude of vibration is involved. In order to find the amount of this damping we calculate the quantity of energy which flows out through a spherical surface with the oscillator at the center, in accordance with the law of Poynting. However, we may not place the energy flowing outwards in accordance with this law through the spherical surface in an infinitely small interval of time dt equal to the energy radiated in the same time interval from the oscillator. For, in general, the electromagnetic energy does not always flow in the outward direction, but flows alternately outwards and inwards, and we should obtain in this manner for the quantity of the radiation outwards, values which are alternately positive and negative, and which also depend essentially upon the radius of the supposed sphere in such manner

that they increase toward infinity with decreasing radius—
which is opposed to the fundamental conception of radiated
energy. This energy will, moreover, be only found independent of the radius of the sphere when we calculate the total
amount of energy flowing outwards through the surface of the
sphere, not for the time element dt, but for a sufficiently large
time. If the vibrations are purely periodic, we may choose for
the time a period; if this is not the case, which for the sake
of generality we must here assume, it is not possible to specify a priori any more general criterion for the least possible
necessary magnitude of the time than that which makes the
energy radiated essentially independent of the radius of the
supposed sphere.

In this way we succeed in finding for the energy emitted
from the oscillator in the time from t to $t + \mathfrak{T}$ the following
expression:

$$\frac{2}{3c^3} \int_t^{t+\mathfrak{T}} \ddot{f}^2(t)dt.$$

If now, the oscillator be in an electromagnetic field which has
the electric component $\mathfrak{E}_z$ at the oscillator in the direction
of its axis, then the energy absorbed by the oscillator in the
same time is:

$$\int_t^{t+\mathfrak{T}} \mathfrak{E}_z \dot{f} \cdot dt.$$

Hence, the principle of conservation of energy is expressed in
the following form:

$$\int_t^{t+\mathfrak{T}} \left(\frac{dU}{dt} + \frac{2}{3c^3} \ddot{f}^2 - \mathfrak{E}_z \dot{f} \right) dt = 0.$$

This equation, together with the assumption that the constant

$$\frac{4\pi^2 \nu_0}{3c^3 L} = \sigma \tag{42}$$

is a small number, leads to the following linear differential equation for the vibrations of the oscillator:

$$K f + L \ddot{f} - \frac{2}{3c^3} \dddot{f} = \mathfrak{E}_z. \tag{43}$$

In accordance with what precedes, in so far as the oscillator is excited into vibrations by an external field $\mathfrak{E}_z$, one may designate it as a resonator which possesses the natural period ν_0 and the small logarithmic decrement σ. The same equation may be obtained from the electron theory, but I have considered it an advantage to derive it in a manner independent of any hypothesis concerning the nature of the resonator.

Now, let the resonator be in a vacuum filled with stationary black radiation of specific intensity $\mathfrak{K}_\nu$. How, then, does the mean energy U of the resonator in a state of stationary vibration depend upon the specific intensity of radiation $\mathfrak{K}_{\nu_0}$ with the natural period ν_0 of the corresponding color? It is this question which we have still to consider today. Its answer will be found by expressing on the one hand the energy of the resonator U and on the other hand the intensity of radiation $\mathfrak{K}_{\nu_0}$ by means of the component $\mathfrak{E}_z$ of the electric field exciting the resonator. Now however complicated this quantity may be, it is capable of development in any case for a very large time interval, from $t = 0$ to $t = \mathfrak{T}$, in the Fourier's series:

$$\mathfrak{E}_z = \sum_{n=1}^{n=\infty} C_n \cos\left(\frac{2\pi n t}{\mathfrak{T}} - \vartheta_n\right), \tag{44}$$

and for this same time interval $\mathfrak{T}$ the moment of the resonator in the form of a Fourier's series may be calculated as a function of t from the linear differential equation (43). The initial condition of the resonator may be neglected if we only consider such times t as are sufficiently far removed from the origin of time $t = 0$.

If it be now recalled that in a stationary state of vibration the mean energy U of the resonator is given, in accordance with (40), (41) and (42), by:

$$U = K\bar{f}^2 = \frac{16\pi^4 \nu_0{}^3}{3\sigma c^3}\bar{f}^2,$$

it appears after substitution of the value of f obtained from the differential equation (43) that:

$$U = \frac{3c^3}{64\pi^2\nu_0{}^2}\mathfrak{T}\bar{C}_{n0}{}^2, \qquad (45)$$

wherein $\bar{C}_{n0}{}^2$ denotes the mean value of C_n for all the series of numbers n which lie in the neighborhood of the value $\nu_0\mathfrak{T}$, i. e., for which $\nu_0\mathfrak{T}$ is approximately $= 1$.

Now let us consider on the other hand the intensity of black radiation, and for this purpose proceed from the space density of the total radiation. In accordance with (30), this is:

$$\epsilon = \frac{8\pi}{c}\int_0^\infty \mathfrak{K}_\nu d\nu = \frac{1}{8\pi}(\bar{\mathfrak{E}}_x{}^2 + \bar{\mathfrak{E}}_y{}^2 + \bar{\mathfrak{E}}_z{}^2 + \bar{\mathfrak{H}}_x{}^2 + \bar{\mathfrak{H}}_y{}^2 + \bar{\mathfrak{H}}_z{}^2),$$

$$(46)$$

and therefore, since the radiation is isotropic, in accordance with (44):

$$\frac{8\pi}{c}\int_0^\infty \mathfrak{K}_\nu d\nu = \frac{3}{4\pi}\bar{\mathfrak{E}}_z{}^2 = \frac{3}{8\pi}\sum_{n=1}^{n=\infty} C_n{}^2.$$

If we write $\Delta n/\mathfrak{T}$ on the left instead of $d\nu$, where Δn is a large number, we get:

$$\frac{8\pi}{c}\sum_{n=1}^{n=\infty} \mathfrak{K}_\nu \frac{\Delta n}{\mathfrak{T}} = \frac{3}{8\pi}\sum_{n=1}^{n=\infty} C_n{}^2,$$

and obtain then by "spectral" division of this equation:

$$\frac{8\pi}{c}\mathfrak{K}_{\nu_0}\frac{\Delta n}{\mathfrak{T}} = \frac{3}{8\pi}\sum_{n_0-(\Delta n/2)}^{n_0+(\Delta n/2)} C_n{}^2,$$

and, if we introduce again the mean value

$$\frac{1}{\Delta n} \cdot \sum_{n_0-(\Delta n/2)}^{n_0+(\Delta n/2)} C_n{}^2 = \bar{C}_{n0}{}^2,$$

we then get:

$$\mathfrak{K}_{\nu_0} = \frac{3c\mathfrak{T}}{64\pi^2} \cdot \bar{C}_{n0}.$$

By comparison with (45) the relation sought is now found:

$$\mathfrak{K}_{\nu_0} = \frac{\nu_0{}^2}{c^2} U, \tag{47}$$

which is striking on account of its simplicity and, in particular, because it is quite independent of the damping constant σ of the resonator.

This relation, found in a purely electrodynamic manner, between the spectral intensity of black radiation and the energy of a vibrating resonator will furnish us in the next lecture, with the aid of thermodynamic considerations, the necessary means of attack in deriving the temperature of black radiation together with the distribution of energy in the normal spectrum.

6 HEAT RADIATION. STATISTICAL THEORY

Following the preparatory considerations of the last lecture we shall treat today the problem which we have come to recognize as one of the most important in the theory of heat radiation: the establishment of that universal function which governs the energy distribution in the normal spectrum. The means for the solution of this problem will be furnished us through the calculation of the entropy S of a resonator placed in a vacuum filled with black radiation and thereby excited into stationary vibrations. Its energy U is then connected with the corresponding specific intensity $\mathfrak{K}_\nu$ and its natural frequency ν in the radiation of the surrounding field through equation (47):

$$\mathfrak{K}_\nu = \frac{\nu^2}{c^2} U. \tag{48}$$

When S is found as a function of U, the temperature T of the resonator and that of the surrounding radiation will be given by:

$$\frac{dS}{dU} = \frac{1}{T}, \tag{49}$$

and by elimination of U from the last two equations, we then find the relationship among $\mathfrak{K}_\nu$, T and ν.

In order to find the entropy S of the resonator we will utilize the general connection between entropy and probability, which we have extensively discussed in the previous lectures, and inquire then as to the existing probability that the vibrating resonator possesses the energy U. In accordance with what we have seen in connection with the elucidation of the second law through atomistic ideas, the second law is only applicable to a physical system when we consider the quantities which determine the state of the system as mean values of numerous disordered individual values, and the probability of a state is then equal to the number of the numerous, a priori equally probable, complexions which make possible the realization of the state. Accordingly, we have to consider the energy U of a resonator placed in a stationary field of black radiation as a constant mean value of many disordered independent individual values, and this procedure agrees with the fact that every measurement of the intensity of heat radiation is extended over an enormous number of vibration periods. The entropy of a resonator is then to be calculated from the existing probability that the energy of the radiator possesses a definite mean value U within a certain time interval.

In order to find this probability, we inquire next as to the existing probability that the resonator at any fixed time possesses a given energy, or in other words, that that point (the state point) which through its coordinates indicates the state of the resonator falls in a given "state domain." At the conclusion of the third lecture (p. 92) we saw in general that this probability is simply measured through the magnitude of the corresponding state domain:

$$\int d\varphi \cdot d\psi,$$

in case one employs as coordinates of state the general coordinate φ and the corresponding momentum ψ. Now in general, the energy of the resonator, in accordance with (40), is:

$$U = \tfrac{1}{2}Kf^2 + \tfrac{1}{2}L\dot{f}^2.$$

If we choose f as the general coordinate φ and put, therefore, $\varphi = f$, then the corresponding impulse ψ is equal

$$\frac{\partial U}{\partial \dot{f}} = L\dot{f},$$

and the energy U expressed as a function of φ and ψ is:

$$U = \tfrac{1}{2}K\varphi^2 + \frac{1}{2}\frac{\psi^2}{L}.$$

If now we desire to find the existing probability that the energy of a resonator shall lie between U and $U + \Delta U$, we have to calculate the magnitude of that state domain in the (φ, ψ)-plane which is bounded by the curves $U = $ const. and $U + \Delta U = $ const. These two curves are similar and similarly placed ellipses and the portion of surface bounded by them is equal to the difference of the areas of the two ellipses. The areas are respectively U/ν and $(U + \Delta U)/\nu$; consequently, the magnitude sought for the state domain is: $\Delta U/\nu$. Let us now consider the whole state plane so divided into elementary portions by a large number of ellipses, such that the annular areas between consecutive ellipses are equal to each other; i.e., so that:

$$\frac{\Delta U}{\nu} = \text{const} = h.$$

We thus obtain those portions ΔU of the energy which correspond to equal probabilities and which are therefore to be designated as the energy elements:

$$\epsilon = \Delta U = h\nu. \tag{50}$$

If the determination of the elementary domains is effected in a manner quite similar to that employed in the kinetic gas theory, there exist, with respect to the relationships there found, very notable differences. In the first place, the state of the physical system considered here, the resonator, does not depend as there upon the coordinates and the velocities, but upon the energy only, and this circumstance necessitates that the entropy of a state depend, not upon the distribution of the state quantities φ and ψ, but only upon the energy U. A further difference consists in this, that we have to do in the case of molecules with spacial mean values, but in the case of radiation with mean values as regards time. But this distinction may be disregarded when we reflect that the mean time value of the energy U of a given resonator is obviously identical with the mean space value at a given instant of time of a great number N of similar resonators distributed in the same stationary field of radiation. Of course these resonators must be placed sufficiently far apart in order not directly to influence one another. Then the total energy of all the resonators:

$$U_N = NU \qquad (51)$$

is quite irregularly distributed among all the individual resonators, and we have referred back the disorder as regards time to a disorder as regards space.

We are now concerned with the probability W of the state determined by the energy U_N of the N resonators placed in the same stationary field of radiation; i.e., with the number of individual arrangements or complexions which correspond to the distribution of energy U_N among the N resonators. With this in view, we subdivide the given total energy U_N into its elements ϵ so that:

$$U_N = P\epsilon. \qquad (52)$$

These P energy elements are to be distributed in every possible manner among the N resonators. Let us consider, then, the N resonators to be numbered and the figures written beside one another in a series, and in such manner that the number of times each figure appears is equal to the number of energy elements which fall upon the corresponding resonator. Then we obtain through such a number series a representation of a fixed complexion, in which with each individual resonator there is associated a definite energy. For example, if there are $N = 4$ resonators and $P = 6$ energy elements present, then one of the possible complexions is represented by the number series

$$1 \quad 1 \quad 3 \quad 3 \quad 3 \quad 4$$

which asserts that the first resonator contains two, the second 0, the third 3, and the fourth 1 energy element. The totality of numbers in the series is 6, equal to the number of the energy elements present. The arrangement of figures in the series is immaterial for any complexion, since the mere interchange of figures does not change the energy of a given resonator. The number of all the possible different complexions is therefore equal to the number of possible "combinations with repetition" of 4 elements with 6 classes:

$$W = \frac{(4 + 6 - 1)!}{(4 - 1)!\, 6!} = \frac{9!}{3!\, 6!} = 84,$$

or, in our general case the probability sought is:

$$W = \frac{(N + P - 1)!}{(N - 1)!\, P!}.$$

We obtain, therefore, for the entropy S_N of the resonator system, in accordance with equation (12), since N and P are large numbers,

$$S_N = k \log \frac{(N + P)!}{N!\, P!}$$

and with the aid of Sterling's formula (16):

$$S_N = k\{(N + P) \log(N + P) - N \log N - P \log P\}.$$

If, in accordance with (52), we now write U_N/ϵ for P, NU for U_N in accordance with (51), and $h\nu$ for ϵ, in accordance with (50), we obtain, after an easy transformation, for the mean entropy of a single resonator:

$$\frac{S_N}{N} = S = k\left\{\left(1 + \frac{U}{h\nu}\right)\log\left(1 + \frac{U}{h\nu}\right) - \frac{U}{h\nu}\log\frac{U}{h\nu}\right\}$$

as the solution of the problem in hand.

We will now introduce the temperature T of the resonator, and will express through T the energy U of the resonator and also the intensity $\mathfrak{K}_\nu$ of the heat radiation related to it through a stationary state of energy exchange. For this purpose we utilize equation (49) and obtain then for the energy of the resonator:

$$U = \frac{h\nu}{e^{h\nu/kT} - 1}.$$

It is to be observed that we have not here to do with a uniform distribution of energy (cf. p. 104) among the various resonators.

For the specific intensity of the monochromatic plane polarized ray of frequency ν, we have, in accordance with (48):

$$\mathfrak{K}_\nu = \frac{h\nu^3}{c^2} \cdot \frac{1}{e^{h\nu/kT} - 1}. \tag{53}$$

This expression furnishes for each temperature T the energy distribution in the normal spectrum of a black body. A comparison with equation (38) of the last lecture furnishes us then with the universal function:

$$F(\nu, T) = \frac{h\nu^3}{e^{h\nu/kT} - 1}.$$

If we refer the specific intensity of a monochromatic ray, not to the frequency ν, but, as is commonly done in experimental physics, to the wave length λ, then, since between the absolute values of $d\nu$ and $d\lambda$ the relation exists:

$$|d\nu| = \frac{c \cdot |d\lambda|}{\lambda^2},$$

we obtain from

$$E_\lambda |d\lambda| = \mathfrak{K}_\nu |d\nu|,$$

the relation:

$$E_\lambda = \frac{c^2 h}{\lambda^5} \cdot \frac{1}{e^{ch/k\lambda T} - 1} \tag{54}$$

as the intensity of a monochromatic plane polarized ray of wave length λ is emitted normally to the surface of a black body in a vacuum at temperature T. For small values of λT (54) reduces to:

$$E_\lambda = \frac{c^2 h}{\lambda^5} \cdot e^{-(ch/k\lambda T)}, \tag{55}$$

which expresses Wien's Displacement Law. For large values of λT on the other hand, there results from (54):

$$E_\lambda = \frac{ckT}{\lambda^4}, \tag{56}$$

a relation first established by Lord Rayleigh and which we may here designate as the Rayleigh Law of Radiation.

From equation (30), taking account of (53), we obtain for the space density of black radiation in a vacuum:

$$\epsilon = \frac{48\pi h}{c^3} \left(\frac{kT}{h}\right)^4 \cdot \alpha = aT^4,$$

wherein

$$\alpha = 1 + \frac{1}{2^4} + \frac{1}{3^4} + \frac{1}{4^4} + \cdots = 1.0823.$$

The Stefan-Boltzmann law is hereby expressed. In accordance with the measurements of Kurlbaum, we have the constant

$$a = \frac{48\pi k^4}{c^3 h^3} \cdot \alpha = 7.061 \cdot 10^{-15} \frac{\text{erg}}{\text{cm}^3 \text{deg}^4}.$$

For that wave length λ_m which corresponds in the spectrum of black radiation to the maximum intensity of radiation E_λ we have from equation (54):

$$\left(\frac{dE_\lambda}{d\lambda}\right)_{\lambda=\lambda_m} = 0.$$

Carrying out the differentiation, we get, after putting for brevity:

$$\frac{ch}{k\lambda_m T} = \beta, \quad e^{-\beta} + \frac{\beta}{5} - 1 = 0.$$

The root of this transcendental equation is:

$$\beta = 4.9651;$$

and $\lambda_m T = ch/k\beta = b$ is a constant (Wien's Displacement Law). In accordance with the measurements of O. Lummer and E. Pringsheim,

$$b = 0.294 \text{ cm} \cdot \text{deg}.$$

From this there follow the numerical values

$$k = 1.346 \times 10^{-16} \frac{\text{erg}}{\text{deg}}, \quad \text{and} \quad h = 6.548 \times 10^{-27} \text{erg} \cdot \text{sec}.$$

The value found for k easily permits of the specification numerically, in the C.G.S. system, of the general connection between entropy and probability, as expressed through the universal equation (12). Thus, quite in general, the entropy of a physical system is:

$$S = 1.346 \times 10^{-16} \overset{e}{\log} W.$$

In the application to the kinetic gas theory we obtain from equation (24) for the ratio of the molecular mass to the mol mass:

$$\omega = \frac{k}{R} = 1.62 \times 10^{-24},$$

i.e., to one mol there corresponds $1/\omega = 6.175 \times 10^{23}$ molecules, where it is supposed that the mol of oxygen

$$O_2 = 32\text{g}.$$

Accordingly, the number of molecules contained in 1 cu. cm. of an ideal gas at $0°$ Cels. and at atmospheric pressure is:

$$N = 2.76 \times 10^{19}.$$

The mean kinetic energy of the progressive motion of a molecule at the absolute temperature $T = 1$ in the absolute C.G.S. system, in accordance with (27), is:

$$L = \tfrac{3}{2}k = 2.02 \times 10^{-16}.$$

In general, the mean kinetic energy of progressive motion of a molecule is expressed by the product of this number and the absolute temperature T.

The elementary quantum of electricity, or the free electric charge of a monovalent ion or electron, in electrostatic measure is:

$$e = \omega \times 9658 \times 3 \times 10^{10} = 4.69 \times 10^{-10}.$$

This result stands in noteworthy agreement with the results of the latest direct measurements of the electric elementary quantum made by E. Rutherford and H. Geiger, and E. Regener.

Even if the radiation formula (54) here derived had shown itself as valid with respect to all previous tests, the theory would still require an extension as regards a certain point; for in it the physical meaning of the universal constant h remains quite unexplained. All previous attempts to derive a radiation formula upon the basis of the known laws of electron theory, among which the theory of J. H. Jeans is to be considered as the most general and exact, have led to the conclusion that

h is infinitely small, so that, therefore, the radiation formula of Rayleigh possesses general validity, but, in my opinion, there can be no doubt that this formula loses its validity for short waves, and that the pains which Jeans has taken to place[1] the blame for the contradiction between theory and experiment upon the latter are unwarranted.

Consequently, there remains only the one conclusion, that previous electron theories suffer from an essential incompleteness which demands a modification, but how deeply this modification should go into the structure of the theory is a question upon which views are still widely divergent. J. J. Thompson inclines to the most radical view, as do J. Larmor, A. Einstein, and with him I. Stark, who even believe that the propagation of electromagnetic waves in a pure vacuum does not occur precisely in accordance with the Maxwellian field equations, but in definite energy quanta $h\nu$. I am of the opinion, on the other hand, that at present it is not necessary to proceed in so revolutionary a manner, and that one may come successfully through by seeking the significance of the energy quantum $h\nu$ solely in the mutual actions with which the resonators influence one another.[2] A definite decision with regard to these important questions can only be brought about as a result of further experience.

---

[1] In that the walls used in the measurements of hollow space radiations must be diathermanous for the shortest waves.

[2] It is my intention to give a complete presentation of these relations in Volume 31 of the Annalen der Physik.

7 GENERAL DYNAMICS.
PRINCIPLE OF LEAST ACTION

Since I began three weeks ago today to depict for you the present status of the system of theoretical physics and its probable future development, I have continually sought to bring out that in the theoretical physics of the future the most important and the final division of all physical processes would likely be into reversible and irreversible processes. In succeeding lectures, with the aid of the calculus of probability and with the introduction of the hypothesis of elementary disorder, we have seen that all irreversible processes may be considered as reversible elementary processes: in other words, that irreversibility does not depend upon an elementary property of a physical process, but rather depends upon the ensemble of numerous disordered elementary processes of the same kind, each one of which individually is completely reversible, and upon the introduction of the macroscopic method of treatment. From this standpoint one can say quite correctly that in the final analysis all processes in nature are reversible. That there is herein contained no contradiction to the principle regarding the irreversibility of processes expressed in terms of the mean values of elementary processes of macroscopic changes of state, I have demonstrated fully in the third lecture. Perhaps it will be appropriate at this place to interject a more general statement. We are accustomed in physics to seek the explanation of a natural process by the method of division of the process into elements. We regard each complicated process as composed of simple elementary processes, and seek to analyse it through thinking of the whole as the sum of the parts. This method, however, presupposes that through this division the character of the whole is not changed; in somewhat similar manner each measurement of a physical process presupposes that the progress of the phenomena is not influenced by the introduction of the measuring instrument. We have here a case in which that supposition is not warranted, and where a direct conclusion with regard to the parts applied to the whole leads to quite false results. If we divide an irreversible process into its elementary constituents, the disorder and along with it the irreversibility vanishes; an irreversible process must remain beyond the understanding of anyone who relies upon the fundamental law: that all properties of the whole must also be recognizable in the parts. It appears to me as though a similar difficulty presents itself in most of the problems of intellectual life.

Now after all the irreversibility in nature thus appears in a certain sense eliminated, it is an illuminating fact that general elementary dynamics has only to do with reversible processes. Therefore we shall occupy ourselves in what follows with reversible processes exclusively. That which makes this procedure so valuable for the theory is the circumstance that all known reversible processes, be they mechanical, electrodynamical or thermal, may be brought together under a single principle which answers unambiguously all questions regarding their behavior. This principle is not that of conservation of energy; this holds, it is true, for all these processes, but does not determine unambiguously their behavior; it is the more comprehensive principle of least action.

The principle of least action has grown upon the ground of mechanics where it enjoys equal rank and regard with numerous other principles; the principle of d'Alembert, the principle

of virtual displacement, Gauss's principle of least constraint,
the Lagrangian Equations of the first and second kind. All
these principles are equivalent to one another and therefore at
bottom are only different formularizations of the same laws;
sometimes one and sometimes another is the most convenient
to use. But the principle of least action has the decided advantage over all the other principles mentioned in that it connects
together in a single equation the relations between quantities
which possess, not only for mechanics, but also for electrodynamics and for thermodynamics, direct significance, namely,
the quantities: space, time and potential. This is the reason
why one may directly apply the principle of least action to
processes other than mechanical, and the result has shown
that such applications, as well in electrodynamics as in thermodynamics, lead to the appropriate laws holding in these
subjects. Since a representation of a unified system of theoretical physics such as we have here in mind must lay the
chief emphasis upon as general an interpretation as possible
of physical laws, it is self evident that in our treatment the
principle of least action will be called upon to play the principal rôle. I desire now to show how it is applied in simple
individual cases.

The general formularization of the principle of least action
in the interpretation given to it by Helmholz is as follows:
among all processes which may carry a certain arbitrarily
given physical system subject to given external actions from
a given initial position into a given final position in a given
time, the process which actually takes place in nature is that
which is distinguished by the condition that the integral

$$\int_{t_0}^{t_1} (\delta H + A)dt = 0, \tag{57}$$

wherein an arbitrary displacement of the independent coordinates (and velocities) is denoted by the sign δ, and A denotes
the infinitely small increase in energy (external work) which

the system experiences in the displacement δ. The function H is the kinetic potential. When we speak here of the positions, the coordinates, and the velocities of the configuration, we understand thereby, not only those special ones corresponding to mechanical ideas, but also all the so-called generalized coordinates with the quantities derived therefrom; and these may represent equally well quantities of electricity, volumes, and the like.

In the applications which we shall now make of the principle of least action, we must first decide as to whether the generalized coordinates which determine the state of the system considered are present in finite number or form a continuous infinite manifold. We shall distinguish the examples here considered in accordance with this viewpoint.

7.1 The Position (Configuration) is Determined by a Finite Number of Coordinates

In ordinary mechanics this is actually the case in every system of a finite number of material points or rigid bodies among whose coordinates there exist arbitrary fixed equations of condition. If we call the independent coordinates φ_1, φ_2, $\cdots$, then the external work is:

$$A = \Phi_1 \delta\varphi_1 + \Phi_2 \delta\varphi_2 + \cdots = \delta E, \tag{58}$$

wherein Φ_1, Φ_2, $\cdots$ are the "external force components" which correspond to the individual coordinates, and E denotes the energy of the system. Then the principle of least action is expressed by:

$$\int_{t_0}^{t_1} dt \cdot \sum_{1,2,\cdots} \left(\frac{\partial H}{\partial \varphi_1} \delta\varphi_1 + \frac{\partial H}{\partial \dot{\varphi}_1} \delta\dot{\varphi}_1 + \Phi_1 \delta\varphi_1 \right) = 0.$$

From this follow the equations of motion:

$$\Phi_1 - \frac{d}{dt}\left(\frac{\partial H}{\partial \dot{\varphi}_1}\right) + \frac{\partial H}{\partial \varphi_1} = 0, \tag{59}$$

and so on for all the indices, 1, 2, $\cdots$. Through multiplication of the individual equations by $\dot{\varphi}_1$, $\dot{\varphi}_2$, $\cdots$ addition and integration with respect to time, there results the equation of conservation of energy, whereby the energy E is given by the expression:

$$E = \sum_{1,2,\cdots} \dot{\varphi}_1 \frac{\partial H}{\partial \dot{\varphi}_1} - H. \tag{60}$$

In ordinary mechanics $H = L - U$, if L denote the kinetic and U the potential energy. Since L is a homogeneous function of the second degree with respect to the $\dot{\varphi}$'s, it follows from (60) that:

$$E = 2L - H = L + U.$$

But this expression holds by no means in general.

We pass now to the consideration of the quasi-stationary motion of a system of linear conductors carrying simple closed galvanic currents. The state of the system is given by the position and the velocities of the conductors and by the current densities in each of the same. The coordinates referring to the position of the first conductor may be represented by φ_1, φ_1', φ_1'', $\cdots$, corresponding designations holding for the remaining conductors. We inquire now as to the increase of energy or the external work, A, which corresponds to a virtual displacement of all coordinates. Energy may be conveyed to the system through mechanical actions and through electromagnetic induction as well. The former corresponds to mechanical work, the latter to electromotive work. The former will be of the familiar form:

$$\Phi_1 \delta\varphi_1 + \Phi_1' \delta\varphi_1 + \cdots + \Phi_2 \delta\varphi_2 + \cdots.$$

If we denote by E_1, E_2, $\cdots$ the electromotive forces which are induced in the individual conductors through external agencies (e.g., moving magnets which do not belong to the system), then the electromotive work done from outside upon the currents in the conductors of the system is:

$$E_1\delta\epsilon_1 + E_2\delta\epsilon_2 + \cdots ,$$

if $\delta\epsilon_1$, $\delta\epsilon_2$, $\cdots$ denote the quantities of electricity which pass through cross sections of the conductors due to infinitely small virtual currents. The finite current densities will then be denoted by $\dot{\epsilon}_1$, $\dot{\epsilon}_2$, $\cdots$. The electrical state of the first conductor is thus determined in general by the current density $\dot{\epsilon}_1$, the mechanical state (position and velocity) by the coordinates φ_1, $\varphi_1{}'$, $\varphi_1{}''$, $\cdots$ and the corresponding velocities $\dot{\varphi}_1$, $\dot{\varphi}_1'$, $\dot{\varphi}_1''$, $\cdots$. The coordinates ϵ_1, ϵ_2, $\cdots$ are so-called "cyclical" coordinates, since the state does not depend upon their momentary values, but only upon their differential quotients with respect to time, just as, for example, the state of a body rotatable about an axis of symmetry depends only upon the angular velocity, and not upon the angle of rotation. The scheme of notation adopted permits of the direct application of the above formularization of the principle of least action to the case here considered. Thus $H = H_\phi + H_\epsilon$, where H_ϕ, the mechanical potential, depends only upon the φ's and $\dot{\varphi}$'s, while the electrokinetic potential H_ϵ takes the following form:

$$H_\epsilon = \tfrac{1}{2}L_{11}\dot{\epsilon}_1{}^2 + L_{12}\dot{\epsilon}_1\dot{\epsilon}_2 + L_{13}\dot{\epsilon}_1\dot{\epsilon}_3 + \cdots + \tfrac{1}{2}L_{22}\dot{\epsilon}_2{}^2 + \cdots .$$

The quantities L_{11}, L_{12}, L_{13} $\cdots$ L_{22}, $\cdots$ the coefficients of self induction and mutual induction depend, however, in a definite manner upon the coordinates of position φ_1, $\varphi_1{}'$, $\varphi_1{}''$, $\cdots$, φ_2, $\varphi_2{}'$, $\varphi_2{}''$, $\cdots$.

In accordance with (59), we have for the motion of the first conductor:

$$\Phi_1 - \frac{d}{dt}\left(\frac{\partial H_\phi}{\partial \dot{\varphi}_1}\right) + \frac{\partial H_\phi}{\partial \varphi_1} + \frac{\partial H_\epsilon}{\partial \varphi_1} = 0,$$

with corresponding equations for φ_1', φ_1'', $\cdots$, and for the electric current in it:

$$E_1 - \frac{d}{dt}\left(\frac{\partial H_\epsilon}{\partial \dot{\epsilon}_1}\right) = 0.$$

The laws for the mechanical (ponderomotive) actions may be condensed into the statement that, in addition to the ordinary force upon the first conductor expressed by Φ_1, there is a mechanical force

$$\frac{\partial H_\epsilon}{\partial \varphi_1} = \frac{1}{2}\frac{\partial L_{11}}{\partial \varphi_1}\dot{\epsilon}_1{}^2 + \frac{\partial L_{12}}{\partial \varphi_1}\dot{\epsilon}_1\dot{\epsilon}_2 + \frac{\partial L_{13}}{\partial \varphi_1}\dot{\epsilon}_1\dot{\epsilon}_3 + \cdots,$$

which is composed of an action of the current upon itself (first term) and of the actions of the remaining currents upon it (following terms).

The laws of electrical action, on the other hand, are expressed by the statement, that to the external electromotive force E_1 in the first conductor there is added the electromotive force

$$-\frac{d}{dt}\left(\frac{\partial H_\epsilon}{\partial \dot{\epsilon}_1}\right) = -\frac{d}{dt}(L_{11}\dot{\epsilon}_1 + L_{12}\dot{\epsilon}_2 + L_{13}\dot{\epsilon}_3 + \cdots)$$

which likewise is composed of an action of the current upon itself (self induction) and of the inducing actions of the remaining currents, and that these two forces compensate each other.

The galvanic conductance or the galvanic resistance is not contained in these equations because the corresponding energy, Joule heat, is produced in an irreversible manner, and irreversible processes are not represented by the principle of least action. One can formally include this action, likewise any other irreversible action, in accordance with the procedure of Helmholz, by introducing it as an external force, in the present case as the electromotive force due to the resistance w, which operates to cause a diminution in the energy of the system. For an infinitely small element of time, the amount of this energy change is:

$$-(w_1\dot{\epsilon}_1{}^2+w_2\dot{\epsilon}_2{}^2+w_3\dot{\epsilon}_3{}^2+\cdots)\cdot dt = -(w_1\dot{\epsilon}_1 d\epsilon_1+w_2\dot{\epsilon}_2 d\epsilon_2+\cdots).$$

Consequently, since the external work $E_1 d\epsilon_1 + E_2 d\epsilon_2 + \cdots$ now includes the Joule heat, the external force components $E_1, E_2, \cdots$ in the electromotive equations must be increased by the additional terms $-w_1\dot{\epsilon}_1, -w_2\dot{\epsilon}_2, \cdots$.

The application of the principle of least action to thermodynamic processes is of special interest, because the importance of the question relating to the fixing of the generalized coordinates, which determine the state of the system, here becomes prominent. From the standpoint of pure thermodynamics, the variables which determine the state of a body can certainly be quite arbitrarily chosen, e.g., in the case of a gas of invariable constitution any two of the following quantities may be chosen as independent variables and all others expressed through them: volume V, temperature T, pressure P, energy E, entropy S. In the present case, the matter is quite different. If we inquire, in order to apply the principle of least action, with regard to the energy change or the total work A which will be done upon the gas from without in an infinitely small virtual displacement, it may be written in the form:

$$A = -p \cdot \delta V + T \cdot \delta S.$$

$T\delta S$ is the heat added from without, $-p\delta V$ the mechanical work furnished from without. In order to bring this into agreement with the general formula for external work (58):

$$A = \Phi_1 \delta\varphi_1 + \Phi_2 \delta\varphi_2$$

it becomes necessary now to choose V and S as the generalized coordinates of state and, therefore, to identify with them the previously employed quantities φ_1 and φ_2. Then $-p$ and T are the generalized force components Φ_1 and Φ_2. Now, since in thermodynamics every reversible change of state proceeds with infinite slowness, the velocity components $\dot{V}$ and $\dot{S}$, and in general all differential coefficients with respect to time, are to be placed equal to zero, and the principle of least action (59) reduces to:

$$\Phi + \frac{\partial H}{\partial \varphi} = 0,$$

and, therefore, in our case:

$$-p + \left(\frac{\partial H}{\partial V}\right)_S = 0 \quad \text{and} \quad T + \left(\frac{\partial H}{\partial S}\right)_V = 0.$$

Further, in accordance with (60):

$$E = -H.$$

Now these equations are actually valid, since they only present other forms of the relation

$$dS = \frac{dE + pdV}{T}.$$

The view here presented is fundamentally that which is given in the energetics of Mach, Ostwald, Helm, and Wiedeburg. The generalized coordinates V and S are in this theory the "capacity factors," $-p$ and T the "intensity factors."[1] So long as one limits himself to an irreversible process, nothing stands in the way of carrying out this method completely, nor of a generalization to include chemical processes.

In opposition to it there is an essentially different method of regarding thermodynamic processes, which in its complete

---

[1]The breaking up of the energy differentials into two factors by the exponents of energetics is by no means associated with a special property of energy, but is simply an expression for the elementary law that the differential of a function $F(x)$ is equal to the product of the differential dx by the derivative $\dot{F}(x)$.

generality was first introduced into physics by Helmholtz. In accordance with this method, one generalized coordinate is V, and the other is not S, but a certain cyclical coordinate—we shall denote it, as in the previous example, by ϵ—which does not appear itself in the expression for the kinetic potential H and only appears through its differential coefficient, $\dot{\epsilon}$; and this differential coefficient is the temperature T. Accordingly, H is dependent only upon V and T. The equation for the total external work, in accordance with (58), is:

$$A = -p\delta V + E\delta\epsilon,$$

and agreement with thermodynamics is obviously found if we set:

$$E\delta\epsilon = T\delta S, \quad \text{and also:} \quad Ed\epsilon = TdS, \quad Edt = dS.$$

The equations (59) for the principle of least action become:

$$-p + \left(\frac{\partial H}{\partial V}\right)_T = 0 \quad \text{and} \quad E - \frac{d}{dt}\left(\frac{\partial H}{\partial T}\right)_V = 0,$$

or

$$d\left(\frac{\partial H}{\partial T}\right)_V = Edt = dS,$$

or by integration:

$$\left(\frac{\partial H}{\partial T}\right)_V = S,$$

to an additive constant, which we may set equal to 0. For the energy there results, in accordance with (60):

$$E = \dot{\epsilon}\frac{\partial H}{\partial \dot{\epsilon}} - H = T\left(\frac{\partial H}{\partial T}\right)_V - H,$$

and consequently:

$$H = -(E - TS).$$

H is therefore equal to the negative of the function which Helmholz has called the "free energy" of the system, and the above equations are known from thermodynamics.

Furthermore, the method of Helmholz permits of being carried through consistently, and so long as one limits himself to the consideration of reversible processes, it is in general quite impossible to decide in favor of the one method or the other. However, the method of Helmholz possesses a distinct advantage over the other which I desire to emphasize here. It lends itself better to the furtherance of our endeavor toward the unification of the system of physics. In accordance with the purely energetic method, the independent variables V and S have absolutely nothing to do with each other; heat is a form of energy which is distinguished in nature from mechanical energy and which in no way can be referred back to it. In accordance with Helmholz, heat energy is reduced to motion, and this certainly indicates an advance which is to be placed, perhaps, upon exactly the same footing as the advance which is involved in the consideration of light waves as electromagnetic waves.

To be sure, the view of Helmholz is not broad enough to include irreversible processes; with regard to this, as we have earlier stated in detail, the introduction of the calculus of probability is necessary in order to throw light on the question. At the same time, this is also the real reason that the exponents of energetics will have nothing to do with the strict observance of irreversible processes, and they either declare them as doubtful or ignore them completely. In reality, the facts of the case are quite the reverse; irreversible processes are the only processes occurring in nature. Reversible processes form only an ideal abstraction, which is very valuable for the theory, but which is never completely realized in nature.

7.2 The Generalized Coordinates of State Form a Continuous Manifold

The laws of infinitely small motions of perfectly elastic bodies furnish us with the simplest example. The coordinates of state are then the displacement components, $\mathfrak{v}_x$, $\mathfrak{v}_y$, $\mathfrak{v}_z$, of a material point from its position of equilibrium (x, y, z), considered as a function of the coordinates x, y, z. The external work is given by a surface integral:

$$A = \int d\sigma (X_\nu \delta\mathfrak{v}_x + Y_\nu \delta\mathfrak{v}_y + Z_\nu \delta\mathfrak{v}_z)$$

($d\sigma$, surface element; ν, inner normal). The kinetic potential is again given by the difference of the kinetic energy L and the potential energy U:

$$H = L - U.$$

The kinetic energy is:

$$L = \int \frac{d\tau\, k}{2} (\dot{\mathfrak{v}}_x^2 + \dot{\mathfrak{v}}_y^2 + \dot{\mathfrak{v}}_z^2),$$

wherein $d\tau$ denotes a volume element, k the volume density. The potential energy U is likewise a space integral of a homogeneous quadratic function f which specifies the potential energy of a volume element. This depends, as is seen from purely geometrical considerations, only upon the 6 "strain coefficients:"

$$\frac{\partial \mathfrak{v}_x}{\partial x} = x_x, \quad \frac{\partial \mathfrak{v}_y}{\partial y} = y_y, \quad \frac{\partial \mathfrak{v}_z}{\partial z} = z_z,$$

$$\frac{\partial \mathfrak{v}_y}{\partial z} + \frac{\partial \mathfrak{v}_z}{\partial y} = y_z = z_y, \quad \frac{\partial \mathfrak{v}_z}{\partial x} + \frac{\partial \mathfrak{v}_x}{\partial z} = z_x = x_z,$$

$$\frac{\partial \mathfrak{v}_x}{\partial y} + \frac{\partial \mathfrak{v}_y}{\partial x} = x_y = y_x.$$

In general, therefore, the function f contains 21 independent constants, which characterize the whole elastic behavior of the substance. For isotropic substances these reduce on grounds of symmetry to 2. Substituting these values in the expression for the principle of least action (57) we obtain:

$$\int dt \left\{ \int d\tau k(\dot{\mathfrak{v}}_x \delta \dot{\mathfrak{v}}_x + \cdots) - \int d\tau \left(\frac{\partial f}{\partial x_x} \delta x_x + \frac{\partial f}{\partial x_y} \delta x_y + \cdots \right) \right.$$
$$\left. + \int d\sigma (X_\nu \delta \mathfrak{v}_x + \cdots) \right\} = 0.$$

If we put for brevity:

$$-\frac{\partial f}{\partial x_x} = X_x, \qquad -\frac{\partial f}{\partial y_y} = Y_y, \qquad -\frac{\partial f}{\partial z_z} = Z_z,$$

$$-\frac{\partial f}{\partial y_z} = Y_z = Z_y, \quad -\frac{\partial f}{\partial z_x} = Z_x = X_z, \quad -\frac{\partial f}{\partial x_y} = X_y = Y_x,$$

it turns out, as the result of purely mathematical operations in which the variations $\delta \dot{\mathfrak{v}}_x$, $\delta \dot{\mathfrak{v}}_y$, $\cdots$ and likewise the variations δx_x, δx_y, $\cdots$ are reduced through suitable partial integration with respect to the variations $\delta \mathfrak{v}_x$, $\delta \mathfrak{v}_y$, $\cdots$, that the conditions within the body are expressed by:

$$k\ddot{\mathfrak{v}}_x + \frac{\partial X_x}{\partial x} + \frac{\partial X_y}{\partial y} + \frac{\partial X_z}{\partial z} = 0, \quad \cdots$$

and at the surface, by:

$$X_\nu = X_x \cos \nu x + X_y \cos \nu y + X_z \cos \nu z, \quad \cdots$$

as is known from the theory of elasticity. The mechanical significance of the quantities X_x, Y_y, $\cdots$ as surface forces follows from the surface conditions.

For the last application of the principle of least action we will take a special case of electrodynamics, namely, electrodynamic processes in a homogeneous isotropic non-conductor at rest, e.g., a vacuum. The treatment is analogous to that carried out in the foregoing example. The only difference lies in the fact that in electrodynamics the dependence of the potential energy U upon the generalized coordinate $\mathfrak{v}$ is somewhat different than in elastic phenomena.

We therefore again put for the external work:

$$A = \int d\sigma (X_\nu \delta\mathfrak{v}_x + Y_\nu \delta\mathfrak{v}_y + Z_\nu \delta\mathfrak{v}_z), \qquad (61)$$

and for the kinetic potential:

$$H = L - U,$$

wherein again:

$$L = \int d\tau \frac{k}{2}(\dot{\mathfrak{v}}_x{}^2 + \dot{\mathfrak{v}}_y{}^2 + \dot{\mathfrak{v}}_z{}^2) = \int d\tau \frac{k}{2}(\dot{\mathfrak{v}})^2.$$

On the other hand, we write here:

$$U = \int d\tau \frac{h}{2}(\operatorname{curl} \mathfrak{v})^2.$$

Through these assumptions the dynamical equations including the boundary conditions are now completely determined. The principle of least action (57) furnishes:

$$\int dt \{\int d\tau k(\dot{\mathfrak{v}}_x \delta\dot{\mathfrak{v}}_x + \cdots) - \int d\tau h(\operatorname{curl}_x \mathfrak{v}\delta \operatorname{curl}_x \mathfrak{v} + \cdots)$$
$$+ \int d\sigma (X_\nu \delta\mathfrak{v}_x + \cdots)\} = 0.$$

From this follow, in quite an analogous way to that employed above in the theory of elasticity, first, for the interior of the non-conductor:

$$k\ddot{\mathfrak{v}}_x = h \left(\frac{\partial \operatorname{curl}_y \mathfrak{v}}{\partial z} - \frac{\partial \operatorname{curl}_z \mathfrak{v}}{\partial y} \right), \quad \cdots$$

or more briefly

$$k\ddot{\mathfrak{v}} = -h \operatorname{curl} \operatorname{curl} \mathfrak{v}, \qquad (62)$$

and secondly, for the surface:

$$X_\nu = h(\mathrm{curl}_z\, \mathfrak{v} \cdot \cos\nu y - \mathrm{curl}_y\, \mathfrak{v} \cdot \cos\nu z), \quad \cdots \qquad (63)$$

These equations are identical with the known electrodynamical equations, if we identify L with the electric, and U with the magnetic energy (or conversely). If we put

$$L = \frac{1}{8\pi}\int d\tau \cdot \epsilon\mathfrak{E}^2 \quad \text{and} \quad U = \frac{1}{8\pi}\int d\tau \cdot \mu\mathfrak{H}^2,$$

($\mathfrak{E}$ and $\mathfrak{H}$, the field strengths, ϵ, the dielectric constant, μ, the permeability) and compare these values with the above expressions for L and U we may write:

$$\dot{\mathfrak{v}} = -\mathfrak{E}\cdot\sqrt{\frac{\epsilon}{4\pi k}}, \quad \mathrm{curl}\,\mathfrak{v} = \mathfrak{H}\sqrt{\frac{\mu}{4\pi h}}. \qquad (64)$$

It follows then, by elimination of $\mathfrak{v}$, that:

$$\dot{\mathfrak{H}} = -\sqrt{\frac{\epsilon h}{\mu k}}\cdot \mathrm{curl}\,\mathfrak{E},$$

and further, by substitution of $\dot{\mathfrak{v}}$ and $\mathrm{curl}\,\mathfrak{v}$ in equation (62) found above for the interior of the non-conductor, that:

$$\dot{\mathfrak{E}} = \sqrt{\frac{\mu h}{\epsilon k}}\,\mathrm{curl}\,\mathfrak{H}.$$

Comparison with the known electrodynamical equations expressed in Gaussian units:

$$\mu\dot{\mathfrak{H}} = -c\,\mathrm{curl}\,\mathfrak{E}, \quad \epsilon\dot{\mathfrak{E}} = c\,\mathrm{curl}\,\mathfrak{H}$$

(c, velocity of light in vacuum) results in a complete agreement, if we put:

$$\frac{c}{\mu} = \sqrt{\frac{\epsilon h}{\mu k}} \quad \text{and} \quad \frac{c}{\epsilon} = \sqrt{\frac{\mu h}{\epsilon k}}.$$

From either of these two equations it follows that:

$$\frac{h}{k} = \frac{c^2}{\epsilon\mu},$$

the square of the velocity of propagation.

We obtain from (61) for the energy entering the system from without:

$$dt \cdot \int d\sigma (X_\nu \dot{\mathfrak{v}}_x + Y_\nu \dot{\mathfrak{v}}_y + Z_\nu \dot{\mathfrak{v}}_z),$$

or, taking account of the surface equation (63):

$$dt \cdot \int d\sigma h\{(\mathrm{curl}_z\, \mathfrak{v} \cos \nu y - \mathrm{curl}_y\, \mathfrak{v} \cos \nu z)\dot{\mathfrak{v}}_x + \cdots\},$$

an expression which, upon substitution of the values of $\dot{\mathfrak{v}}$ and $\mathrm{curl}\,\mathfrak{v}$ from (64), turns out to be identical with the Poynting energy current.

We have thus by an application of the principle of least action with a suitably chosen expression for the kinetic potential H arrived at the known Maxwellian field equations.

Are, then, the electromagnetic processes thus referred back to mechanical processes? By no means; for the vector $\mathfrak{v}$ employed here is certainly not a mechanical quantity. It is moreover not possible in general to interpret $\mathfrak{v}$ as a mechanical quantity, for instance, $\mathfrak{v}$ as a displacement, $\dot{\mathfrak{v}}$ as a velocity, $\mathrm{curl}\,\mathfrak{v}$ as a rotation. Thus, e.g., in an electrostatic field $\dot{\mathfrak{v}}$ is constant. Therefore, $\mathfrak{v}$ increases with the time beyond all limits, and $\mathrm{curl}\,\mathfrak{v}$ can no longer signify a rotation.[2] While from these considerations the possibility of a mechanical explanation of electrical phenomena is not proven, it does appear, on the other hand, to be undoubtedly true that the significance

---

[2]With regard to the impossibility of interpreting electrodynamic processes in terms of the motions of a continuous medium, cf. particularly, H. Witte: "Über den gegenwärtigen Stand der Frage nach einer mechanischen Erklärung der elektrischen Erscheinungen" Berlin, 1906 (E. Ebering).

of the principle of least action may be essentially extended
beyond ordinary mechanics and that this principle can therefore also be utilized as the foundation for general dynamics,
since it governs all known reversible processes.

8 GENERAL DYNAMICS.
PRINCIPLE OF RELATIVITY

In the lecture of yesterday we saw, by means of examples, that all continuous reversible processes of nature may be represented as consequences of the principle of least action, and that the whole course of such a process is uniquely determined as soon as we know, besides the actions which are exerted upon the system from without, the kinetic potential H as a function of the generalized coordinates and their differential coefficients with respect to time. The determination of this function remains then as a special problem, and we recognize here a rich field for further theories and hypotheses. It is my purpose to discuss with you today an hypothesis which represents a magnificent attempt to establish quite generally the dependency of the kinetic potential H upon the velocities, and which is commonly designated as the principle of relativity. The gist of this principle is: it is in no wise possible to detect the motion of a body relative to empty space; in fact, there is absolutely no physical sense in speaking of such a motion. If, therefore, two observers move with uniform but different velocities, then each of the two with exactly the same right may assert that with respect to empty space he is at rest, and there are no physical methods of measurement enabling us to decide in favor of the one or the other. The principle of relativity in its generalized form is a very recent development. The preparatory steps were taken by H. A. Lorentz, it was first generally formulated by A. Einstein, and was developed into a finished mathematical system by H. Minkowski. However, traces of it extend quite far back into the past, and therefore it seems desirable first to say something concerning the history of its development.

The principle of relativity has been recognized in mechanics since the time of Galilee and Newton. It is contained in the form of the simple equations of motion of a material point, since these contain only the acceleration and not the velocity of the point. If, therefore, we refer the motion of the point, first to the coordinates x, y, z, and again to the coordinates x', y', z' of a second system, whose axes are directed parallel to the first and which moves with the velocity ν in the direction of the positive x-axis:

$$x' = x - \nu t, \quad y' = y, \quad z' = z, \tag{65}$$

and the form of the equations of motion is not changed in the slightest. Nothing short of the assumption of the general validity of the relativity principle in mechanics can justify the inclusion by physics of the Copernican cosmical system, since through it the independence of all processes upon the earth of the progressive motion of the earth is secured. If one were obliged to take account of this motion, I should have, e.g., to admit that the piece of chalk in my hand possesses an enormous kinetic energy, corresponding to a velocity of something like 30 kilometers per second.

It was without doubt his conviction of the absolute validity of the principle of relativity which guided Heinrich Hertz in the establishment of his fundamental equations for the electrodynamics of moving bodies. The electrodynamics of Hertz is, in fact, wholly built upon the principle of relativity. It recognizes no absolute motion with regard to empty space. It speaks only of motions of material bodies relative to one another. In accordance with the theory of Hertz, all electrodynamic processes occur in material bodies; if these move, then the electrodynamic processes occurring therein move with them. To speak of an independent state of motion of a medium outside of material bodies, such as the ether, has just as little sense in the theory of Hertz as in the modern theory of relativity.

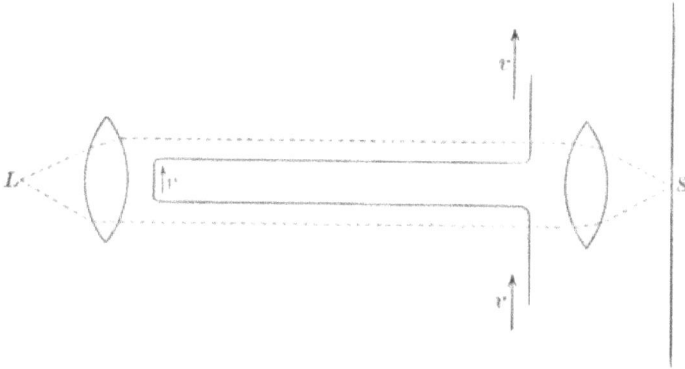

Fig. 2

But the theory of Hertz has led to various contradictions with experience. I will refer here to the most important of these. Fizeau brought (1851) into parallelism a bundle of rays originating in a light source L by means of a lens and then brought it to a focus by means of a second lens upon a screen S (Fig. 2). In the path of the parallel light rays between the two lenses he placed a tube system of such sort that a transparent liquid could be passed through it, and in such manner that in one half (the upper) the light rays would pass in the direction of flow of the liquid while in the other half (the lower), the rays would pass in the opposite direction.

If now a liquid or a gas flow through the tube system with the velocity ν, then, in accordance with the theory of Hertz, since light must be a process in the substance, the light waves must be transported with the velocity of the liquid. The velocity of light relative to L and S is, therefore, in the upper part $q_0 + \nu$, and the lower part $q_0 - \nu$, if q_0 denote the velocity of light relative to the liquid. The difference of these two velocities, 2ν, should be observable at S through corresponding interference of the lower and the upper light rays, and quite independently of the nature of the flowing substance. Experiment did not confirm this conclusion. Moreover, it showed in gases generally no trace of the expected action; i. e., light is propagated in a flowing gas in the same manner as in a gas at rest. On the other hand, in the case of liquids an effect was certainly indicated, but notably smaller in amount than that demanded by the theory of Hertz. Instead of the expected velocity difference 2ν, the difference $2\nu(1 - 1/n^2)$ only was observed, where n is the refractive index of the liquid. The factor $(1 - 1/n^2)$ is called the Fresnel coefficient. There is contained (for $n = 1$) in this expression the result obtained in the case of gases.

It follows from the experiment of Fizeau that, as regards electrodynamic processes in a gas, the motion of the gas is practically immaterial. If, therefore, one holds that electrodynamic processes require for their propagation a substantial carrier, a special medium, then it must be concluded that this medium, the ether, remains at rest when the gas moves in an arbitrary manner. This interpretation forms the basis of the electrodynamics of Lorentz, involving an absolutely quiescent ether. In accordance with this theory, electrodynamic phenomena have only indirectly to do with the motion of matter. Primarily all electrodynamical actions are propagated in ether at rest. Matter influences the propagation only in a secondary way, so far as it is the cause of exciting in greater or less degree resonant vibrations in its smallest parts by means of the electrodynamic waves passing through it. Now, since the refractive properties of substances are also influenced through the resonant vibrations of its smallest particles, there results from this theory a definite connection between the refractive index and the coefficient of Fresnel, and this connection is, as calculation shows, exactly that demanded by measurements. So far, therefore, the theory of Lorentz is confirmed through experience, and the principle of relativity is divested of its general significance.

The principle of relativity was immediately confronted by a new difficulty. The theory of a quiescent ether admits the idea of an absolute velocity of a body, namely the velocity relative to the ether. Therefore, in accordance with this theory, of two observers A and B who are in empty space and who move relatively to each other with the uniform velocity v, it would be at best possible for only one rightly to assert that he is at rest relative to the ether. If we assume, e.g., that at the moment at which the two observers meet an instantaneous optical signal, a flash, is made by each, then an infinitely thin spherical wave spreads out from the place of its origin in all directions through empty space. If, therefore, the observer A remain at the center of the sphere, the observer B will not remain at the center and, as judged by the observer B, the light in his own direction of motion must travel (with the velocity $c - v$) more slowly than in the opposite direction (with the velocity $c + v$), or than in a perpendicular direction (with the velocity $\sqrt{c^2 - v^2}$) (cf. Fig. 3). Under suitable conditions the observer B should be able to detect and measure this sort of effect.

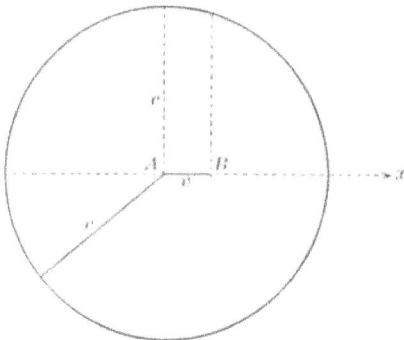

Fig. 3

This elementary consideration led to the celebrated attempt of Michelson to measure the motion of the earth relative to the ether. A parallel beam of rays proceeding from L (Fig. 4) falls upon a transparent plane parallel plate P inclined at $45°$, by which it is in part transmitted and in part reflected. The transmitted and reflected beams are brought into interference by reflection from suitable metallic mirrors S_1 and S_2, which are removed by the same distance l from P. If, now, the earth with the whole apparatus moves in the direction PS_1 with the velocity v, then the time which the light needs in order to go from P to S_1 and back is:

$$\frac{l}{c-v} + \frac{l}{c+v} = \frac{2l}{c}\left(1 + \frac{v^2}{c^2} + \cdots\right).$$

On the other hand, the time which the light needs in order to pass from P to S_2 and back to P is:

$$\frac{l}{\sqrt{c^2-v^2}} + \frac{l}{\sqrt{c^2-v^2}} = \frac{2l}{c}\left(1 + \frac{1}{2}\frac{v^2}{c^2} + \cdots\right).$$

If, now, the whole apparatus be turned through a right angle,

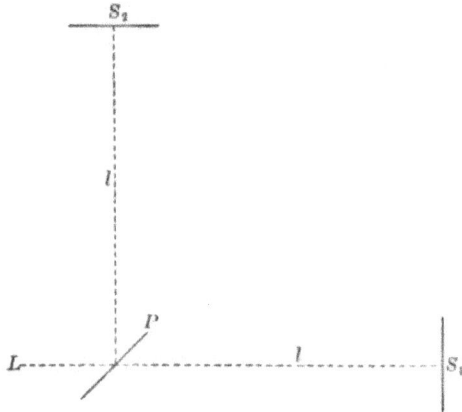

Fig. 4

a noticeable displacement of the interference bands should

result, since the time for the passage over the path PS_2 is now longer. No trace was observed of the marked effect to be expected.

Now, how will it be possible to bring into line this result, established by repeated tests with all the facilities of modern experimental art? E. Cohn has attempted to find the necessary compensation in a certain influence of the air in which the rays are propagated. But for anyone who bears in mind the great results of the atomic theory of dispersion and who does not renounce the simple explanation which this theory gives for the dependence of the refractive index upon the color, without introducing something else in its place, the idea that a moving absolutely transparent medium, whose refractive index is absolutely $= 1$, shall yet have a notable influence upon the velocity of propagation of light, as the theory of Cohn demands, is not possible of assumption. For this theory distinguishes essentially a transparent medium, whose refractive index is $= 1$, from a perfect vacuum. For the former the velocity of propagation of light in the direction of the velocity ν of the medium with relation to an observer at rest is

$$q = c + \frac{\nu^2}{c},$$

for a vacuum, on the other hand, $q = c$. In the former medium, Cohn's theory of the Michelson experiment predicts no effect, but, on the other hand, the Michelson experiment should give a positive effect in a vacuum.

In opposition to E. Cohn, H. A. Lorentz and FitzGerald ascribe the necessary compensation to a contraction of the whole optical apparatus in the direction of the earth's motion of the order of magnitude ν^2/c^2. This assumption allows better of the introduction again of the principle of relativity, but it can first completely satisfy this principle when it appears, not as a necessary hypothesis made to fit the present special case, but as a consequence of a much more general postulate.

We have to thank A. Einstein for the framing of this postulate and H. Minkowski for its further mathematical development.

Above all, the general principle of relativity demands the renunciation of the assumption which led H. A. Lorentz to the framing of his theory of a quiescent ether; the assumption of a substantial carrier of electromagnetic waves. For, when such a carrier is present, one must assume a definite velocity of a ponderable body as definable with respect to it, and this is exactly that which is excluded by the relativity principle. Thus the ether drops out of the theory and with it the possibility of mechanical explanation of electrodynamic processes, i. e., of referring them to motions. The latter difficulty, however, does not signify here so much, since it was already known before, that no mechanical theory founded upon the continuous motions of the ether permits of being completely carried through (cf. p. 150). In place of the so-called free ether there is now substituted the absolute vacuum, in which electromagnetic energy is independently propagated, like ponderable atoms. I believe it follows as a consequence that no physical properties can be consistently ascribed to the absolute vacuum. The dielectric constant and the magnetic permeability of a vacuum have no absolute meaning, only relative. If an electrodynamic process were to occur in a ponderable medium as in a vacuum, then it would have absolutely no sense to distinguish between field strength and induction. In fact, one can ascribe to the vacuum any arbitrary value of the dielectric constant, as is indicated by the various systems of units. But how is it now with regard to the velocity of propagation of light? This also is not to be regarded as a property of the vacuum, but as a property of electromagnetic energy which is present in the vacuum. Where there is no energy there can exist no velocity of propagation.

With the complete elimination of the ether, the opportunity is now present for the framing of the principle of relativity. Obviously, we must, as a simple consideration shows, introduce something radically new. In order that the moving observer B mentioned above (Fig. 3, p. 157) shall not see the light signal given by him travelling more slowly in his own direction of motion (with the velocity $c - \nu$) than in the opposite direction (with the velocity $c + \nu$), it is necessary that he shall not identify the instant of time at which the light has covered the distance $c - \nu$ in the direction of his own motion with the instant of time at which the light has covered the distance $c + \nu$ in the opposite direction, but that he regard the latter instant of time as later. In other words: the observer B measures time differently from the observer A. This is a priori quite permissible; for the relativity principle only demands that neither of the two observers shall come into contradiction with himself. However, the possibility is left open that the specifications of time of both observers may be mutually contradictory.

It need scarcely be emphasized that this new conception of the idea of time makes the most serious demands upon the capacity of abstraction and the projective power of the physicist. It surpasses in boldness everything previously suggested in speculative natural phenomena and even in the philosophical theories of knowledge: non-euclidean geometry is child's play in comparison. And, moreover, the principle of relativity, unlike non-euclidean geometry, which only comes seriously into consideration in pure mathematics, undoubtedly possesses a real physical significance. The revolution introduced by this principle into the physical conceptions of the world is only to be compared in extent and depth with that brought about by the introduction of the Copernican system of the universe.

Since it is difficult, on account of our habitual notions concerning the idea of absolute time, to protect ourselves, without special carefully considered rules, against logical mistakes in the necessary processes of thought, we shall adopt the mathematical method of treatment. Let us consider then an electrodynamic process in a pure vacuum; first, from the standpoint of an observer A; secondly, from the standpoint of an observer B, who moves relatively to observer A with a velocity ν in the direction of the x-axis. Then, if A employ the system of reference x, y, z, t, and B the system of reference x', y', z', t', our first problem is to find the relations among the primed and the unprimed quantities. Above all, it is to be noticed that since both systems of reference, the primed and the unprimed, are to be like directed, the equations of transformation between corresponding quantities in the two systems must be so established that it is possible through a transformation of exactly the same kind to pass from the first system to the second, and conversely, from the second back to the first system. It follows immediately from this that the velocity of light c' in a vacuum for the observer B is exactly the same as for the observer A. Thus, if c' and c are different, $c' > c$, say, it would follow that: if one passes from one observer A to another observer B who moves with respect to A with uniform velocity, then he would find the velocity of propagation of light for B greater than for A. This conclusion must likewise hold quite in general independently of the direction in which B moves with respect to A, because all directions in space are equivalent for the observer A. On the same grounds, in passing from B to A, c must be greater than c', for all directions in space for the observer B are now equivalent. Since the two inequalities contradict, therefore c' must be equal to c. Of course this important result may be generalized immediately, so that the totality of the quantities independent of the motion, such as the velocity of light in a vacuum, the constant of gravitation between two bodies at rest, every isolated electric charge, and the entropy of any physical system possess the same values for both observers. On the other hand, this law does not hold for quantities such as energy, volume, temperature, etc. For these quantities depend also upon the velocity, and a body which is at rest for A

is for B a moving body.

We inquire now with regard to the form of the equations of transformation between the unprimed and the primed coordinates. For this purpose let us consider, returning to the previous example, the propagation, as it appears to the two observers A and B, of an instantaneous signal creating an infinitely thin light wave which, at the instant at which the observers meet, begins to spread out from the common origin of coordinates. For the observer A the wave travels out as a spherical wave:

$$x^2 + y^2 + z^2 - c^2t^2 = 0. \tag{66}$$

For the second observer B the same wave also travels as a spherical wave with the same velocity:

$$x'^2 + y'^2 + z'^2 - c^2t'^2 = 0; \tag{67}$$

for the first observer has no advantage over the second ob-

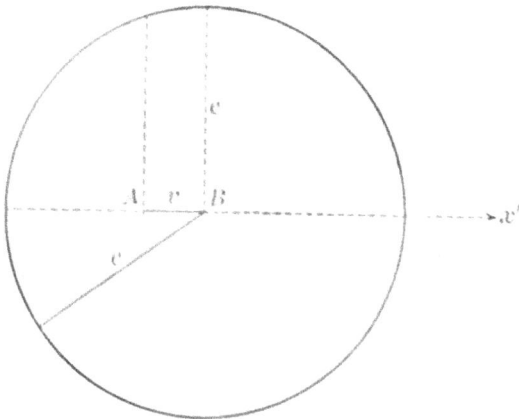

Fig. 5

server. B can exactly with the same right as A assert that he is at rest at the center of the spherical wave, and for B, after unit time, the wave appears as in Fig. 5, while its appearance for the observer A after unit time, is represented by Fig. 3 (p. 157).[1]

The equations of transformation must therefore fulfill the condition that the two last equations, which represent the same physical process, are compatible with each other; and furthermore: the passage from the unprimed to the primed quantities must in no wise be distinguished from the reverse passage from the primed to the unprimed quantities. In order to satisfy these conditions, we generalize the equations of transformation (65), set up at the beginning of this lecture for the old mechanical principle of relativity, in the following manner:

$$x' = \kappa(x - \nu t), \quad y' = \lambda y, \quad z' = \mu z, \quad t' = \nu t + \rho x.$$

Here ν denotes, as formerly, the velocity of the observer B relative to A and the constants κ, λ, μ, ν, ρ are yet to be determined. We must have:

$$x = \kappa'(x' - \nu' t'), \quad y = \lambda' y', \quad z = \mu' z', \quad t = \nu' t' + \rho' x'.$$

It is now easy to see that λ and λ' must both $= 1$. For, if, e.g., λ be greater than 1, then λ' must also be greater than 1; for the two transformations are equivalent with regard to the y axis. In particular, it is impossible that λ and λ' depend upon the direction of motion of the other observer. But now, since, in accordance with what precedes, $\lambda = 1/\lambda'$, each of the two inequalities contradict and therefore $\lambda = \lambda' = 1$; likewise, $\mu = \mu' = 1$. The condition for identity of the two spherical waves then demands that the expression (66):

$$x^2 + y^2 + z^2 - c^2 t^2$$

---

[1]The circumstance that the signal is a finite one, however small the time may be, has significance only as regards the thickness of the spherical layer and not for the conclusions here under consideration.

become, through the transformation of coordinates, identical with the expression (67):

$$x'^2 + y'^2 + z'^2 - c^2 t'^2,$$

and from this the equations of transformation follow without ambiguity:

$$x' = \kappa(x - vt), \quad y' = y, \quad z' = z, \quad t' = \kappa\left(t - \frac{v}{c^2}x\right), \quad (68)$$

wherein

$$\kappa = \frac{c}{\sqrt{c^2 - v^2}}.$$

Conversely:

$$x = \kappa(x' + vt'), \quad y = y', \quad z = z', \quad t = \kappa\left(t' + \frac{v}{c^2}x'\right). \quad (69)$$

These equations permit quite in general of the passage from the system of reference of one observer to that of the other (H. A. Lorentz), and the principle of relativity asserts that all processes in nature occur in accordance with the same laws and with the same constants for both observers (A. Einstein). Mathematically considered, the equations of transformation correspond to a rotation in the four dimensional system of reference (x, y, z, ict) through the imaginary angle $\operatorname{arctg}(i(v/c))$ (H. Minkowski). Accordingly, the principle of relativity simply teaches that there is in the four dimensional system of space and time no special characteristic direction, and any doubts concerning the general validity of the principle are of exactly the same kind as those concerning the existence of the antipodians upon the other side of the earth.

We will first make some applications of the principle of relativity to processes which we have already treated above. That the result of the Michelson experiment is in agreement with the principle of relativity, is immediately evident; for, in accordance with the relativity principle, the influence of a uniform motion of the earth upon processes on the earth can under no conditions be detected.

We consider now the Fizeau experiment with the flowing liquid (see p. 155). If the velocity of propagation of light in the liquid at rest be again q_0, then, in accordance with the relativity principle, q_0 is also the velocity of the propagation of light in the flowing liquid for an observer who moves with the liquid, in case we disregard the dispersion of the liquid; for the color of the light is different for the moving observer. If we call this observer B and the velocity of the liquid as above, ν, we may employ immediately the above formulae in the calculation of the velocity of propagation of light in the flowing liquid, judged by an observer A at the screen S. We have only to put

$$\frac{dx'}{dt'} = x' = q_0,$$

to seek the corresponding value of

$$\frac{dx}{dt} = \dot{x}.$$

For this obviously gives the velocity sought.

Now it follows directly from the equations of transformation (69) that:

$$\frac{dx}{dt} = \dot{x} = \frac{\dot{x}' + \nu}{1 + \dfrac{\nu \dot{x}'}{c^2}},$$

and, therefore, through appropriate substitution, the velocity sought in the upper tube, after neglecting higher powers in ν/c and ν/q_0, is:

$$\dot{x} = \frac{q_0 + \nu}{1 + \dfrac{\nu q_0}{c^2}} = q_0 + \nu \left(1 - \frac{q_0^2}{c^2}\right),$$

and the corresponding velocity in the lower tube is:

$$q_0 - \nu \left(1 - \frac{q_0^2}{c^2}\right).$$

The difference of the two velocities is

$$2\nu \left(1 - \frac{q_0^2}{c^2}\right) = 2\nu \left(1 - \frac{1}{n^2}\right),$$

which is the Fresnel coefficient, in agreement with the measurements of Fizeau.

The significance of the principle of relativity extends, not only to optical and other electrodynamic phenomena, but also to all processes of ordinary mechanics; but the familiar expression $(\frac{1}{2}mq^2)$ for the kinetic energy of a mass point moving with the velocity q is incompatible with this principle.

But, on the other hand, since all mechanics as well as the rest of physics is governed by the principle of least action, the significance of the relativity principle extends at bottom only to the particular form which it prescribes for the kinetic potential H, and this form, though I will not stop to prove it, is characterized by the simple law that the expression

$$H\Delta dt$$

for every space element of a physical system is an invariant

$$= H'\Delta dt'$$

with respect to the passage from one observer A to the other observer B or, what is the same thing, the expression $H/\sqrt{c^2 - q^2}$ is in this passage an invariant $= H'/\sqrt{c^2 - q'^2}$.

Let us now make some applications of this very general law, first to the dynamics of a single mass point in a vacuum, whose state is determined by its velocity q. Let us call the kinetic potential of the mass point for $q = 0$, H_0, and consider now the point at an instant when its velocity is q. For an observer B who moves with the velocity q with respect to the observer A, $q' = 0$ at this instant, and therefore $H' = H_0$. But now since in general:

$$\frac{H}{\sqrt{c^2 - q^2}} = \frac{H'}{\sqrt{c^2 - q'^2}},$$

we have after substitution:

$$H = \sqrt{1 - \frac{q^2}{c^2}} \Delta H_0 = \sqrt{1 - \frac{\dot{x}^2 + \dot{y}^2 + \dot{z}^2}{c^2}} \Delta H_0.$$

With this value of H, the Lagrangian equations of motion (59) of the previous lecture are applicable.

In accordance with (60), the kinetic energy of the mass point amounts to:

$$E = \dot{x}\frac{\partial H}{\partial \dot{x}} + \dot{y}\frac{\partial H}{\partial \dot{y}} + \dot{z}\frac{\partial H}{\partial \dot{z}} - H = q\frac{\partial H}{\partial q} - H = -\frac{H_0}{\sqrt{1 - \frac{q^2}{c^2}}},$$

and the momentum to:

$$G = \frac{\partial H}{\partial q} = -\frac{qH_0}{c\sqrt{c^2 - q^2}}.$$

G/q is called the transverse mass m_t, and dG/dq the longitudinal mass m_l of the point; accordingly:

$$m_t = -\frac{H_0}{c\sqrt{c^2 - q^2}}, \quad m_l = -\frac{cH_0}{(c^2 - q^2)^{3/2}}.$$

For $q = 0$, we have

$$m_t = m_l = m_0 = -\frac{H_0}{c^2}.$$

It is apparent, if one replaces in the above expressions the constant H_0 by the constant m_0, that the momentum is:

$$G = \frac{m_0 q}{\sqrt{1 - \frac{q^2}{c^2}}}$$

and the transverse mass:

$$m_t = \frac{m_0}{\sqrt{1 - \frac{q^2}{c^2}}},$$

and the longitudinal mass:

$$m_l = \frac{m_0}{\left(1 - \dfrac{q^2}{c^2}\right)^{3/2}},$$

and, finally, that the kinetic energy is:

$$E = \frac{m_0 c^2}{\sqrt{1 - \dfrac{q^2}{c^2}}} = m_0 c^2 + \tfrac{1}{2} m_0 q^2 + \cdots.$$

The familiar value of ordinary mechanics $\frac{1}{2} m_0 q^2$ appears here therefore only as an approximate value. These equations have been experimentally tested and confirmed through the measurements of A. H. Bucherer and of E. Hupka upon the magnetic deflection of electrons.

A further example of the invariance of $H \Delta dt$ will be taken from electrodynamics. Let us consider in any given medium any electromagnetic field. For any volume element V of the medium, the law holds that $V \Delta dt$ is invariant in the passage from the one to the other observer. It follows from this that H/V is invariant; i. e., the kinetic potential of a unit volume or the "*space density of kinetic potential*" is invariant.

Hence the following relation exists;

$$\mathfrak{E}\mathfrak{D} - \mathfrak{H}\mathfrak{B} = \mathfrak{E}'\mathfrak{D}' - \mathfrak{H}'\mathfrak{B}',$$

wherein $\mathfrak{E}$ and $\mathfrak{H}$ denote the field strengths and $\mathfrak{D}$ and $\mathfrak{B}$ the corresponding inductions. Obviously a corresponding law for the space energy density $\mathfrak{E}\mathfrak{D} + \mathfrak{H}\mathfrak{B}$ will not hold.

A third example is selected from thermodynamics. If we take the velocity q of a moving body, the volume V and the temperature T as independent variables, then, as I have shown in the previous lecture (p. 144), we shall have for the pressure p and the entropy S the following relations:

$$\frac{\partial H}{\partial V} = p \quad \text{and} \quad \frac{\partial H}{\partial T} = S.$$

Now since $V/\sqrt{c^2 - q^2}$ is invariant, and S likewise invariant (see p. 162), it follows from the invariance of $H/\sqrt{c^2 - q^2}$ that p is invariant and also that $T/\sqrt{c^2 - q^2}$ is invariant, and hence that:

$$p = p' \quad \text{and} \quad \frac{T}{\sqrt{c^2 - q^2}} = \frac{T'}{\sqrt{c^2 - q'^2}}.$$

The two observers A and B would estimate the pressure of a body as the same, but the temperature of the body as different.

A special case of this example is supplied when the body considered furnishes a black body radiation. The black body radiation is the only physical system whose dynamics (for quasi-stationary processes) is known with absolute accuracy. That the black body radiation possesses inertia was first pointed out by F. Hasenöhrl. For black body radiation at rest the energy $E_0 = aT^4V$ is given by the Stefan-Boltzmann law, and the entropy $S_0 = \int (dE_0/T) = \frac{4}{3}aT^3V$, and the pressure $p_0 = (a/3)T^4$, and, therefore, in accordance with the above relations, the kinetic potential is:

$$H_0 = \frac{a}{3}T^4V.$$

Let us imagine now a black body radiation moving with the velocity q with respect to the observer A and introduce an observer B who is at rest $(q = 0)$ with reference to the black body radiation; then:

$$\frac{H}{\sqrt{c^2 - q^2}} = \frac{H'}{\sqrt{c^2 - q'^2}} = \frac{H_0'}{c},$$

wherein

$$H_0' = \frac{a}{3}T'^4V'.$$

Taking account of the above general relations between T' and T, V' and V, this gives for the moving black body radiation the kinetic potential:

$$H = \frac{a}{3} \frac{T^4 V}{\left(1 - \dfrac{q^2}{c^2}\right)^2},$$

from which all the remaining thermodynamic quantities: the pressure p, the energy E, the momentum G, the longitudinal and transverse masses m_l and m_t of the moving black body radiation are uniquely determined.

Colleagues, ladies and gentlemen, I have arrived at the conclusion of my lectures. I have endeavored to bring before you in bold outline those characteristic advances in the present system of physics which in my opinion are the most important. Another in my place would perhaps have made another and better choice and, at another time, it is quite likely that I myself should have done so. The principle of relativity holds, not only for processes in physics, but also for the physicist himself, in that a fixed system of physics exists in reality only for a given physicist and for a given time. But, as in the theory of relativity, there exist invariants in the system of physics: ideas and laws which retain their meaning for all investigators and for all times, and to discover these invariants is always the real endeavor of physical research. We shall work further in this direction in order to leave behind for our successors where possible—lasting results. For if, while engaged in body and mind in patient and often modest individual endeavor, one thought strengthens and supports us, it is this, that we in physics work, not for the day only and for immediate results, but, so to speak, for eternity.

I thank you heartily for the encouragement which you have given me. I thank you no less for the patience with which you have followed my lectures to the end, and I trust that it may be possible for many among you to furnish in the direction indicated much valuable service to our beloved science.

Part III

Nobel Prize Address

1 THE ORIGIN AND DEVELOPMENT OF THE QUANTUM THEORY

In this lecture I will endeavour to give a general account of the origin of the quantum theory, to sketch concisely its development up to the present, and to point out its immediate significance in physics.

Looking back over the last twenty years to the time when the conception and magnitude of the physical quantum of action first emerged from the mass of experimental facts, and looking back at the long and complicated path which finally led to an appreciation of its importance, the whole history of its development reminds me of the well-proved adage that "to err is human." And all the hard intellectual work of an industrious thinker must often appear vain and fruitless, but that striking occurrences sometimes provide him with an irrefutable proof of the fact that at the end of all his attempts, he does ultimately get one step nearer the truth. An indispensable hypothesis, though it does not guarantee a result, often arises from the pursuit of a definite object, the importance of which is not lessened by initial ill-success.

For me, such an object has, for a long time, been the solution of the problem of the distribution of energy in the normal spectrum of radiant heat. Gustav Kirchhoff showed that, in a space bounded by bodies at equal temperatures, but of arbitrary emissive and absorptive powers, the nature of the heat of radiation is completely independent of the nature of the bodies (1).[1] Later, a universal function was proved to exist, which depended only on temperature and wave length, and was in no way related to the properties peculiar to any substance. The discovery of this remarkable function gave promise of a deeper understanding of the relationship of energy to temperature, which forms the chief problem of thermodynamics, and, therefore, also of all molecular physics. There is no way at present available for obtaining this function but to select from all the various kinds of bodies occurring in Nature any one of known emission and absorption coefficients, and to calculate the heat radiation when the exchange of energy is stationary. According to Kirchhoff's theorem, this must be independent of the constitution of the body.

A body especially suited for this purpose appears to be Heinrich Hertz's oscillator, the laws of emission of which, for a given frequency, have recently been fully developed by Hertz (2). If a number of such oscillators be placed in a space enclosed by reflecting walls, they will exchange energy one with another by taking up or emitting electromagnetic waves, analogous with a sound source and resonators, until finally stationary black radiation, so called, obtains in the enclosure according to Kirchhoff's law. At one time I fostered the hope which seems to us rather naive in these days, that the laws of classical electrodynamics, if applied sufficiently generally, and extended by suitable hypotheses, would be sufficient to explain the essential points of the phenomenon looked for, and to lead to the desired goal. To this end, I first of all developed the laws of emission and absorption of a linear resonator in the widest possible way, in fact, by a roundabout way which I could have avoided by using H. A. Lorentz's electron the-

---

[1]The numbers in parentheses refer to the notes at the end of the lecture.

ory then complete in all fundamental points. But since I did not then fully believe in the electron hypothesis, I preferred to consider the energy flowing across a spherical surface of a certain radius enclosing the resonator. This only deals with phenomena in vacuo, but the knowledge of these is enough to enable us to draw the necessary conclusions about the energy changes of the resonator.

The result of this long series of investigations was the establishment of a general relation between the energy of a resonator of given period and the radiant energy of the corresponding region of the spectrum in the surrounding field when the energy exchange is stationary (3). Some of these investigations could be proved by comparison with available observations, particularly the damping measurements of V. Bjerknes, and this is a verification of the results (4). Thus the remarkable conclusion is reached that the relation does not depend on the nature of the resonator, in particular, not upon its damping coefficient – a very gratifying and welcome circumstance to me, since it allowed the whole problem to be simplified in so far that the energy of radiation could be replaced by the energy of the resonator. Thereby a system with one degree of freedom could be substituted for a complicated system with many degrees of freedom.

Indeed, this result was nothing but a step preparatory to starting on the real problem, which now appeared more formidable. The first attempt at solving the problem miscarried; for my original hope proved false, namely, that the radiation emitted from the resonator would, in some characteristic way, be distinct from the absorbed radiation and thus give a differential equation, by solving which it would be possible to derive a condition for the state of stationary radiation. The resonator only responded to the same rays as it emitted, and was not at all sensitive to neighbouring regions of the spectrum.

My assumption that the resonator could exert a one-sided,

i.e. irreversible, effect on the energy of the surrounding field of radiation, was strongly contradicted by Ludwig Boltzmann (5). His mature experience led him to conclude that, according to the laws of classical mechanics, each phenomenon which I had considered, could operate in exactly the reverse direction. Thus, a spherical wave sent out from a resonator may be reversed and proceed in ever diminishing concentric spheres until it shrinks up at the resonator and is absorbed by it, and causes again the energy previously absorbed to be emitted once more into space in the directions along which it had come. Even if, by introducing suitable limits, I could exclude from the hypothesis of "natural radiation" such singular phenomena as spherical waves travelling inwards, all these analyses show clearly that an essential connecting link is still missing for the complete understanding of the problem.

No other course remained open to me but to attack the problem from the opposite direction, namely, through thermodynamics, with which I felt more familiar. Here I was helped by my previous researches into the second law of thermodynamics, and I straightway conceived the idea of connecting the entropy and not the temperature of the resonator with the energy, indeed, not the entropy itself, but its second differential coefficient with respect to energy, since this has a direct physical meaning for the irreversibility of the exchange of energy between resonator and radiation. Since at that time I did not see my way clear to go any further into the dependence of entropy and probability, I could, first of all, only refer to results that had already been obtained. Now, in 1899, the most interesting result was the law of energy distribution which had just been discovered by W. Wien (6). The experimental proof of this was undertaken by F. Paschen at the *Hochschule*, Hanover, and by O. Lummer and E. Pringsheim at the *Reichsanstalt*, Charlottenburg. This law represents the dependence of the intensity of radiation on temperature by means of an exponential function. Using this law to calculate the relation between the entropy and energy of a resonator, the remarkable result is obtained, that R, the reciprocal of the differential coefficient referred to above, is proportional to the energy (7). This exceedingly simple relation is a complete and adequate expression of Wien's law of distribution of energy; for the dependence upon wave length is always given immediately as well as the dependence upon energy by Wien's generally accepted law of displacements (8).

Since the whole problem deals with one of the universal laws of Nature, and since I believed then, as I do now, that the more general a natural law is, the simpler is its form (though it cannot always be said with certainty and finality which is the simpler form), I thought for a long time that the above relation, namely, that R is proportional to the energy, should be considered as the foundation of the law of distribution of energy (9). This idea soon proved to be untenable in the light of more recent results. While Wien's law was confirmed for small values of energy, i.e. for short waves, O. Lummer and E. Pringsheim found large deviations in the case of long waves (10). Finally, the observations made by G. Rubens and F. Kurlbaum, with infra-red rays after transmission through fluorspar and rock salt (11), showed a totally different relation, which, under certain conditions, was still very simple. In this case, R is proportional, not to the energy, but to the square of the energy, and this relation is more accurate the larger the energies and wave lengths considered (12).

Thus, by direct experiment, two simple limits have been fixed for the function R, i.e. for small values of the energy it is proportional to the energy, for large values it is proportional to the square of the energy. It was obvious that in the general case the next step was to express R to the sum of two terms, one involving the first power, the other the second power of the energy, so that the first term was the predominating term for small values of the energy, the second term for large values. This gave a new formula for the radiation (13), which has stood the test of experiment fairly satisfactorily so far. No final exact experimental verification has yet been given and a new proof is badly needed (14).

If, however, the radiation formula should be shown to be absolutely exact, it would possess only a limited value, in the sense that it is a fortunate guess at an interpolation formula. Therefore, since it was first enunciated, I have been trying to give it a real physical meaning, and this problem led me to consider the relation between entropy and probability, along the lines of Boltzmann's ideas. After a few weeks of the most strenuous work of my life, the darkness lifted and an unexpected vista began to appear.

I will digress a little. According to Boltzmann, entropy is a measure of physical probability, and the essence of the second law of thermodynamics is that in Nature, the more often a condition occurs, the more probable it is. In Nature, entropy itself is never measured, but only the difference of entropy, and to this extent one cannot talk of absolute entropy without a certain arbitrariness. Yet, the introduction of an absolute magnitude of entropy, suitably defined, is allowed, since certain general theorems can be expressed very simply by doing so. As far as I can see, it is exactly the same with energy. Energy itself cannot be measured, but only a difference of energy. Therefore, one did not previously deal with energy, but with work, and Ernst Mach, who was concerned to a great extent with the conservation of energy, but avoided all speculations outside the domain of observation, has always refrained from talking of energy itself. Similarly, at first in thermo-chemistry, one considered heat of reaction, i.e. difference of energy, until William Ostwald emphatically showed that many involved considerations could be very much simplified, if one dealt with energy itself instead of calorimetric values. The undetermined additive constant in the expression for energy was fixed later by the relativity theorem of the relation between energy and inertia (15).

As in the case of energy, we can define absolute value for entropy and consequently for physical probability, if the additive constant is fixed so that entropy and energy vanish simultaneously. (It would be better to substitute temperature for energy here.) On this basis a comparatively simple combinatory method was derived for calculating the physical probability of a certain distribution of energy in a system of resonators. This method leads to the same expression for entropy as was obtained from the radiation theory (16). As an offset against much disappointment, I derived much satisfaction from the fact that Ludwig Boltzmann, in a letter acknowledging my paper, gave me to understand that he was interested in, and fundamentally in agreement with, my ideas.

For numerical applications of this method of probability we require two universal constants, each of which has an independent physical significance. The supplementary calculation of these constants from the radiation theory shows whether the method is merely a numerical one or has an actual physical meaning. The first constant is of a more or less formal nature, it depends on the definition of temperature. The value of this constant is 2/3 if temperature be defined as the mean kinetic energy of a molecule in an ideal gas, and is, therefore, a very small quantity (17). With the conventional measure of temperature, however, this constant has an extremely small value, which is naturally closely dependent upon the energy of a single molecule, and an exact knowledge of it leads, therefore, to the calculation of the mass of a molecule and the quantities depending upon it. This constant is frequently called Boltzmann's constant, though Boltzmann himself, to my knowledge, never introduced it – a curious circumstance, explained by the fact that Boltzmann, as appears from various remarks by him (18), never thought of the practicability of measuring this constant exactly. Nothing can better illustrate the impetuous advance made in experimental methods in the last twenty years than the fact that since then, not one only, but a whole series of methods have been devised for measuring the mass of a single molecule with almost the same accuracy as that of a planet.

While, at the time that I carried out the corresponding calculations from the radiation theory, it was impossible to verify exactly the figure obtained, and all that could be achieved was to check the order of magnitude; shortly afterwards, *E*. Rutherford and H. Geiger (19), succeeded in determining the value of the elementary electric charge to be 4.65×10^{-10} electrostatic units, by directly counting α-particles. The agreement of this figure with that calculated by me, 4.69×10^{-10}, was a definite confirmation of the usefulness of my theory. Since then, more perfect methods have been developed by E. Regener, R. A. Millikan, and others (20), and have given a value slightly higher than this.

The interpretation of the second universal constant of the radiation formula was much less simple. I called it the elementary quantum of action, since it is a product of energy and time, and was calculated to be 6.55×10^{-27} erg. sec. Though it was indispensable for obtaining the right expression for entropy – for it is only by the help of it that the magnitude of the standard element of probability could be fixed for the probability calculations (21) – it proved itself unwieldy and cumbrous in all attempts to make it fit in with classical theory in any form. So long as this constant could be considered infinitesimal, as when dealing with large energies or long periods of time, everything was in perfect agreement, but in the general case, a rift appeared, which became more and more pronounced the weaker and more rapid the oscillations considered. The failure of all attempts to bridge this gap soon showed that undoubtedly one of two alternatives must obtain. Either the quantum of action was a fictitious quantity, in which case all the deductions from the radiation theory were largely illusory and were nothing more than mathematical juggling. Or the radiation theory is founded on actual physical ideas, and then the quantum of action must play a fundamental role in physics, and proclaims itself as something quite new and hitherto unheard of, forcing us to recast our physical ideas, which, since the foundation of the infinitesimal calculus by Leibniz and Newton, were built on the assumption of continuity of all causal relations.

Experience has decided for the second alternative. That this decision should be made so soon and so certainly is not due to the verification of the law of distribution of energy in heat radiation, much less to my special derivation of this law, but to the restless, ever advancing labour of those workers who have made use of the quantum of action in their investigations.

The first advance in this work was made by A. Einstein, who proved, on the one hand, that the introduction of the energy quanta, required by the quantum of action, appeared suitable for deriving a simple explanation for a series of remarkable observations of light effects, such as Stokes's rule, emission of electrons, and ionization of gases (22). On the other hand, by identifying the energy of a system of resonators with the energy of a rigid body, he derived a formula for the specific heat of a rigid body, which gives again quite correctly the variation of specific heat, particularly its decrease with decrease of temperature (23). It is not my duty here to give even an approximately complete account of this work. I can only point out the most important characteristic stages in the progress of knowledge.

We will now consider problems in heat and chemistry. As far as the specific heat of a solid body is concerned, Einstein's method, based on the assumption of a single characteristic oscillation of the atom, has been extended by M. Born and Th. von Karman to the case of various characteristic oscillations, more in agreement with practice (24). By greatly simplifying the assumptions regarding the nature of the oscillations, P. Debye obtained a comparatively simple formula for the specific heat of a solid body (25). This not only corroborates, particularly for low temperatures, the experimental values obtained by W. Nernst and his school, but also is in good agreement with the elastic and optical properties of the body. Further, quantum effects are very noticeable when considering the specific heat of gases. W. Nernst had shown at an early stage (26) that the quantum of energy of an oscillation must correspond to the quantum of energy of a rotation, and accordingly expected that the energy of rotation of a gas molecule would decrease with temperature. A. Eucken's measurements of the specific heat of hydrogen verified this deduction (27), and the fact that the calculations of A. Einstein and O. Stern, P. Ehrenfest, and others have not yet been in satisfactory agreement can be ascribed to our incomplete knowledge of the form of the hydrogen molecule. The work of N. Bjerrum, E. v. Bahr, H. Rubens, and G. Hettner, etc., on absorption bands in the infrared rays, shows that there can be no doubt that the rotations of the gas molecules indicated by the quantum conditions do actually exist. However, no one has yet succeeded in giving a complete explanation of these remarkable rotations.

Since all the affinity of a substance is ultimately bound up with its entropy, the theoretical calculation of entropy by means of quanta gives a method of attacking all problems in chemical affinity. Nernst's chemical constant is a characteristic for the absolute value of the entropy of a gas. O. Sackur calculated this constant directly (28) by a combinatory method similar to my method with oscillators, while O. Stern and H. Tetrode, by careful examination of experimental data of evaporation, determined the difference of the entropies of gaseous and non-gaseous substances (29).

The cases considered so far deal with thermodynamical equilibrium, which only give statistical mean values for a number of particles and long periods of time. This observation of electronic impulses, however, leads directly to the dynamical details of the phenomena considered. The determination by J. Franck and G. Hertz of the so called resonance potential, or that critical velocity, the minimum velocity which an electron must have to bring about the emission of a quantum of light by collision with a neutral atom, is as direct a method of measuring the quantum of action as can be desired (30). Also, in the case of the characteristic radiation of the Röntgen spectrum discovered by C. G. Barkla, similar methods which gave very good results were developed by D. L. Webster, E. Wagner, and others.

The liberation of quanta of light by electronic impulses is the converse of the emission of electrons by projection of light, Röntgen or Gamma rays, and here, again, the quanta of energy determined from the quantum of action and the frequency of oscillations play a characteristic part in the same way as we have seen above, in that the velocity of the electrons emitted does not depend on the intensity of the radiation (31), but on the wavelength of the light emitted (32). From a quantitative point of view, also, Einstein's relations for light quanta mentioned above have been verified in every way, particularly by R. A. Millikan, who determined the initial velocities of the emitted electrons (33), while the significance of the light quantum in causing photo chemical reactions has been made clear by E. Warburg (34).

The results quoted above, collected from the most varied branches of physics, present an overwhelming case for the existence of the quantum of action, and the quantum hypothesis was put on a very firm foundation by Niels Bohr's theory of the atom. This theory was destined, by means of the quantum of action, to open a door into the wonderland of spectroscopy, which had obstinately defied all investigators since the discovery of spectral analysis. Once the way was made clear, a mass of new knowledge was obtained concerning this branch of science, as well as allied branches of physics and chemistry. The first brilliant result was Balmer's series for hydrogen and helium, including the reduction of the universal Rydberg constants to pure numbers (35), by which the small difference between hydrogen and helium was found to be due to the slower motion of the heavier atomic core. This led immediately to the investigation of other series in the optical and Röntgen spectra by means of Ritz's useful combination principle, the fundamental meaning of which was now demonstrated for the first time.

In the face of these numerous verifications (which could be considered as very strong proofs in view of the great accuracy of spectroscopic measurements), those who had looked on the problem as a game of chance were finally compelled to throw away all doubt when A. Sommerfeld showed that – by extending the laws of distribution of quanta to systems with several degrees of freedom (and bearing in mind the variability of mass according to the theory of relativity) – an elegant formula follows which must, so far as can be determined by the most delicate measurements now possible (those of F. Paschen (36)), solve the riddle of the structure of hydrogen and helium spectra (37). This is an accomplishment in every way comparable with the famous discovery of the planet Neptune, whose existence and position had been calculated by Leverrier before it had been seen by human eye. Proceeding further along the same lines, P. Epstein succeeded in giving a complete explanation of the Stark effect of the electrical separation of the spectral lines (38), and P. Debye in giving a simple meaning to the K-series of the Röntgen spectrum, investigated by Manne Siegbahn (39). Moreover, there followed a large number of wider investigations, which explained more or less successfully the mystery of the structure of the atom.

In view of all these results – a complete explanation would involve the inclusion of many more well known names – an unbiased critic must recognize that the quantum of action is a universal physical constant, the value of which has been found from several very different phenomena to be 6.54×10^{-27} ergs.

secs. (40). It must seem a curious coincidence that at the time when the idea of general relativity is making headway and leading to unexpected results, Nature has revealed, at a point where it could be least foreseen, an absolute invariable unit, by means of which the magnitude of the action in a time space element can be represented by a definite number, devoid of ambiguity, thus eliminating the hitherto relative character.

Yet no actual quantum theory has been formed by the introduction of the quantum of action. But perhaps this theory is not so far distant as the introduction of Maxwell's light theory was from the discovery of the velocity of light by Olaf Römer. The difficulties in the way of introducing the quantum of action into classical theory from the beginning have been mentioned above. As years have elapsed, these difficulties have increased rather than diminished, and although the impetuous advance of research has dealt with some of them, yet the inevitable gaps remaining in any extension are all the more painful to the conscientious and systematic worker. That which serves as the foundation of the law of action in Bohr's theory is made up of certain hypotheses which were flatly rejected, without any question, a generation ago by physicists. That quite definite orbits determined by quanta are a special feature of the atom may be considered admissible, but it is less easy to assume that the electrons, moving in these paths with a definite acceleration, radiate no energy. But that the quite sharply defined frequency of an emitted light quantum should be different from the frequency of the emitted electrons must seem, at first sight, to a physicist educated in the classical school, an almost unreasonable demand on his imagination.

However, figures are decisive, and the conclusion is that things have been gradually reversed. At first a new foreign element was fitted into a structure, generally considered fixed, with as little change as possible; but now the intruder, after gaining a secure place for itself, has taken the offensive, and today it is almost certain that it will undermine the old structure in some way or other. The question is at what place and to what degree this will happen.

If a surmise be allowed as to the probable outcome of this struggle, everything seems to indicate that the great principles of thermodynamics, derived from the classical theory, will not only maintain their central position in the quantum theory, but will be greatly extended. The adiabatic hypothesis of P. Ehrenfest (41) plays the same part in the quantum theory as the original experiments played in the founding of classical thermodynamics. Just as R. Clausius introduced, as a basis for the measure of entropy, the theorem that any two conditions of a material system are transformable one to the other by reversible processes, so Bohr's new ideas showed the corresponding way to explore the problems opened up by him.

A question, from the complete answer to which we may expect far reaching explanations, is what becomes of the energy of a light quantum after perfect emission? Does it spread out, as it progresses, in all directions, as in Huygens's wave theory, and while covering an ever larger amount of space, diminish without limit? Or does it travel along as in Newton's emanation theory like a projectile in one direction? In the first case the quantum could never concentrate its energy in a particular spot to enable it to liberate an electron from the atomic influences; in the second case we would have the complete triumph of Maxwell's theory, and the continuity between static and dynamic fields must be sacrificed, and with it the present complete explanation of interference phenomena, which have been investigated in all details. Both these alternatives would have very unpleasant consequences for the modern physicist.

In each case there can be no doubt that science will be able to overcome this serious dilemma, and that what seems now to be incompatible may later be regarded as most suitable on account of its harmony and simplicity. Until this goal is attained the problem of the quantum of action will not cease to stimulate research and to yield results, and the greater the difficulties opposed to its solution, the greater will be its significance for the extension and deepening of all our knowledge of physics.

2 Notes

The Bibliography is by no means complete, but serves as an indication of the papers which bear on the subject.

1. G. Kirchhoff. Über das Verhältnis zwischen dem Emissionsvermögen und dem Absorptionsvermögen der Körper fur Warme und Licht. *Gesammelte Abhandlungen.* Leipzig, J. A. Barth, 1882, p. 597 (§17).
2. H. Hertz. *Ann. d. Phys.*, **36**, p. 1, 1889.
3. *Sitz.-Ber. d. Preuss. Akad. d. Wiss.*, 18 May, 1899, p. 455.
4. *Sitz.-Ber. d. Preuss. Akad. d. Wiss.*, 20 Feb., 1896. Also Ann. d. Phys., **60**, p. 577. 1897.
5. L. Boltzmann. *Sitz.-Ber. d. Preuss. Akad. d. Wiss.*, 3 March, 1898, p. 182.
6. W. Wien. *Ann. d. Phys.*, **58**, p. 662, 1896.
7. According to Wien's Law of Energy Distribution, the relation between the energy U of a resonator and the temperature is given by
$$U = a\, e^{-\frac{b}{T}}.$$

If S denotes the entropy, then

$$\frac{1}{T} = \frac{dS}{dU}$$

and R in the text

$$= 1 : \frac{d^2 S}{dU^2} = -b\,U.$$

8. Wien's law of displacements shows that the energy U of a resonator $= \nu f\left(\frac{T}{\nu}\right)$, where ν is the frequency of the oscillations.

9. *Ann. d. Phys.*, **1**, p. 719, 1900.

10. O. Lummer and E. Pringsheim. *Verh. d. Deutsch. Phys. Ges.*, **2**, p. 163, 1900.

11. H. Rubens and F. Kurlbaum. *Sitz.-Ber. der Preuss. Akad. d. Wiss.*, 25 October, 1900, p. 929.

12. For large values of T, the investigations of H. Rubens and F. Kurlbaum give $U = cT$. Then from (7)

$$R = 1 : \frac{d^2 S}{dU^2} = -\frac{U^2}{c}.$$

13. If R be assumed to be equal to

$$1 : \frac{d^2 S}{dU^2} = -bU - \frac{U^2}{c}.$$

then it follows by integration that

$$\frac{1}{T} = \frac{dS}{dU} = \frac{1}{b} \log\left(1 + \frac{bc}{U}\right)$$

and hence the radiation formula

$$U = \frac{bc}{e^{-b/T} - 1}.$$

Cf. *Verh. d. Deutsch. Phys. Ges.*, 19 October, 1900, p. 202.

14. Cf. W. Nernst and Th. Wulf. *Verh. d. Deutsch. Phys. Ges.*, **21**, p. 294, 1919.

15. The absolute value of the energy is equal to the product of the mass and the square of the velocity of light.

16. *Verh. d. Deutsch. Phys. Ges.*, 14 December, 1900, p. 237.

17. In general, if k be the first radiation constant, the mean kinetic energy of a gas molecule

$$U = \tfrac{3}{2}kT.$$

If $T = U$, $k = \frac{2}{3}$. In the case of Kelvin's absolute temperature scale, T is defined by putting the difference of temperature at the boiling and freezing-points of water equal to 100.

18. Cf. e.g. L. Boltzmann. Zur Erinnerung an Josef Loschmidt. *Populäre Schriften*, p. 245, 1905.

19. E. Rutherford and H. Geiger. *Proc. Roy. Soc.*, A., Vol. 81, p. 162, 1908.

20. Cf. R. A. Millikan. *Phys. Zeitschr.*, **14**, p. 796, 1913.

21. The calculation of the probability of a physical state consists of an enumeration of a finite number of equally probable individual cases through which such a state is realized. In order to differentiate these individual cases from one another, it is necessary to fix definitely the nature of each individual case.

22. A. Einstein. *Ann. d. Phys.*, **17**, p. 132, 1905.

23. A. Einstein. *Ann. d. Phys.*, **22**, p. 180, 1907.

24. M. Born and Th. v. Karman. *Phys. Zeitschr.*, **14**, p. 15, 1913.

25. P. Debye. *Ann. d. Phys.*, **39**, p. 789, 1912.

26. W. Nernst. *Phys. Zeitschr.*, **13**, p. 1064, 1912.

27. A. Euchen. *Sitz.-Ber. d. Preuss. Akad. d. Wiss.*, p. 141, 1912.

28. O. Sackur. *Ann. d. Phys.*, **36**, p. 958, 191 1.

29. O. Stern. *Phys. Zeitschr.*, **14**, p. 629, 1913. H. Tetrode. *Proc. Akad. Sci. Amsterdam*, 27 February and 27 March, 1915.

30. J. Franck and G. Hertz. *Verh. d. Deutsch. Phys. Ges.*, **16**, p. 512, 1914.

31. Ph. Lenard. *Ann. d. Phys.*, **8** , p. 149, 1902.

32. E. Ladenburg. *Verh. d. Deutsch. Phys. Ges.*, **9**, p. 504, 1907.

33. R. A. Millikan. *Phys. Zeitschr.*, **17**, p. 217, 1916.

34. E. Warburg, Über den Energieumsatz bei photochemischen Vorgängen in Gasen. *Sitz.-Ber. d. Preuss. Akad. d. Wiss.*, from 1911 onwards.

35. N. Bohr. *Phil. Mag.*, **30**, p. 394, 1915.

36. F. Paschen. *Ann. d. Phys.*, **50**, p. 901, 1916.

37. A. Sommerfeld. *Ann. d. Phys.*, **51**, pp. 1, 125, 1916.

38. P. Epstein. *Ann. d. Phys.*, **50**, p. 489, 1916.

39. P. Debye. *Phys. Zeitschr.*, **18**, p. 276, 1917.

40. E. Wagner. *Ann. d. Phys.*, **57**, p. 467, 1918. R. Ladenburg. Jahr. d. Radioaktivitat u. Elektronik, 17, p. 144, 1920.

41. P. Ehrenfest. *Ann. d. Phys.*, **51**, p. 327, 1916.

www.ingramcontent.com/pod-product-compliance
Lightning Source LLC
Chambersburg PA
CBHW021926190326
41519CB00009B/927